THE FUTURE OF NON-LETHAL WEAPONS

THE FUTURE OF NON-LETHAL WEAPONS

Technologies, Operations, Ethics and Law

Editor

NICK LEWER
University of Bradford

Routledge
Taylor & Francis Group

LONDON AND NEW YORK

First published in 2002 by
FRANK CASS PUBLISHERS

This edition published 2013 by Routledge
2 Park Square, Milton Park, Abingdon, Oxon OX14 4RN
711 Third Avenue, New York, NY 10017

Routledge is an imprint of the Taylor & Francis Group, an informa business

British Library Cataloguing in Publication Data

The future of non-lethal weapons: technologies,
  operations, ethics and law
  1. Nonlethal weapons
  I. Lewer, Nick
  355.8′2

ISBN 0-7146-5309-8 (cloth)
ISBN 0-7146-8265-9 (paper)

Library of Congress Cataloging-in-Publication Data

The future of non-lethal weapons: technologies, operations, ethics, and law /
editor, Nick Lewer.
    p. cm.
  Includes bibliographical references and index.
  ISBN 0-7146-5309-8 (cloth) – ISBN 0-7146-8265-9 (pbk.)
    1. Nonlethal weapons. I. Lewer, Nick.

U795 .F88 2002
623.4–dc21   2002067437

The following chapters first appeared in *Medicine, Conflict and Survival*
(ISSN 1362-3699), Vol.17, No.3 (July–Sept. 2001): 1, 2, 4, 5, 6, 7, 8.
Chapter 3 first appeared in *Peace and Change* (ISSN 0149-0508),
Vol.26, No.1 (2001) and chapter 9 in *Defense Studies* (ISSN 1470-2436),
Vol.1, No.2 (2001).

Contents

Introduction

NICK LEWER

Interest in non-lethal weapons (NLWs), which have been defined as being 'explicitly designed and primarily employed to incapacitate personnel or material while minimising fatalities, permanent injury to personnel, and undesired damage to property and the environment'[1] has increased dramatically during the last ten years[2] as a result of:[3]

- Qualitative advances in non-lethal weapons technology, including dual-use technology applications in civilian/military operations;
- Debates concerning the 'revolution in military affairs',[4] the 'revolution in military technology' and the 'revolution in political affairs';[5]
- A need to find alternatives to lethal methods in peacekeeping operations;
- An increasing role for military forces in operations other than war (OOTW) and military operations in urban terrain (MOUT), including peacekeeping operations;
- Situations in which combatants and non-combatants are mixed together, sometimes deliberately;
- Increasing resistance by domestic constituencies to accept deaths in war operations;
- Debates surrounding inhumane types of NLWs, such as blinding laser weapons;
- Requests from civilian law enforcement agencies and prison services for non-lethal arrest and restraint techniques;
- The concept of being able to fight a 'bloodless and humane' war;
- The presence of international media in war zones and civil disturbances recording the brutality of violent conflict and responses to it.

There is increasing evidence of the use of NLWs in combat and peacekeeping operations. For example, in Kosovo US Marines have fired multi-foam baton rounds, sponge grenades and multi-rubber balls. As one Marine officer reported, 'The bottom line is, despite some issues, the use of non-lethal impact munitions saved lives and kept KFOR [Kosovo Peacekeeping Force] soldiers from having to immediately resort to deadly force'.[6] Also, since the 11 September 2001 atrocities committed by Osama

Bin Laden's al-Qaida network, increased interest has been shown in use of NLWs as anti-terrorist measures on civilian aircraft such as electric stun guns, slippery foam, aerosol sedatives, entangling nets, blinding strobe lights and high-pitched sound.[7]

Emerging Technologies

First- and second-generation non-lethal weapons (which include rubber and plastic bullets, 'bean' bags, electric-shock weapons, incapacitant gases, batons, laser weapons[8] and water cannon) have been described in detail elsewhere.[2] This section will review some of the emerging and developing technologies.

- *Landmines* – The Ottawa Treaty (1997) which banned the use, development, production, stockpiling and transfer of anti-personnel landmines, accelerated research into non-lethal alternatives. A range of mines are now being developed[9] including ones which fire out sticky entanglement nets, electrical stunning wires (a TASER Landmine), small rubber balls (Claymore type) and chemical incapacitants

- *Encapsulation Technology* – is a rapidly developing field which finds particular utility in the delivery of NLWs by enabling their controlled and remote-release, making active materials easier and safer to handle and allowing compartmentalization of multiple component systems (binary chemical weapon systems).[10] Encapsulation can also protect sensitive materials from their environment:[12]
  - In a thermal-activated release, microcapsules could be deployed in front of a target ship and would then be sucked into the ships' water cooling system. Inside the heat exchanger the microcapsule wall would melt, releasing a super adsorbent which would swell and create a gel that blocks the cooling system. The ship would be forced to stop because of overheating or ceased engine.
  - In a pressure-activated release, microcapsules could be deployed in areas to be denied to opposing troops. They could be dropped by air (for example, from unmanned aerial vehicles [UAVs]), or delivered by mortar shells or missiles. When trodden upon, the shell ruptures and releases a malodorant (see below) or other chemical or biological agent.
  - In a chemically-activated release, previously deployed microcapsules can be activated by, for example, water cannon when the shell is dissolved thus releasing the releasing malodorant.

The US Navy is currently researching a 'frangible payload dispensing projectile' which would break open on impact and spray a chemical payload such as CS, CN or OC.[12]

- *Malodorants* – these are also referred to as 'skunk shots' or 'stink bombs' and have potential application for use in riot control and war fighting.[11] Over the years, research has investigated whether such weapons could be ethnically targeted.[12] Some analysts are increasingly worried that malodorants have the potential to violate the Biological and Chemical Weapons Conventions.[13] According to the Sunshine Project, many malodorants can 'be produced by living organisms or are toxins derived from them ... biologically based malodorants that are toxic are unquestionably biological weapons'. Whilst acknowledging that malodorants, compared with pathogen weapons, pose little threat to human life, there is considerable danger that 'malodorant development will encourage lax interpretation and violation' of the Conventions. This concern relating to other biological and chemical weapons that fall within the non-lethal category has also been expressed by Dando.[14]

- *Thermobaric Technology* – for non-lethal personnel incapacitation. The US Navy is examining the feasibility of using thermobaric technology, which produces light, overpressure and heat, to incapacitate humans.[15] This is based on Soviet development of a small heat-plus-pressure explosive which creates a brief but intense fireball that produces a shock wave the Afghans called 'Satan Sticks'.[16]

- *Electromagnetic Directed Energy Weapons* – In early 2001 a new weapon developed in the US was unveiled whose 'aim, in Pentagon speak, is to influence motivational behaviour'.[17] This Vehicle Mounted Active Denial System (VMADS) prompted one reporter to claim that 'not since the advent of gunpowder and the splitting of the atom have armies seen such a leap in technology'.[18] The weapon utilizes part of the electromagnetic spectrum penetrating the skin to a depth of about one-sixty-fourth of an inch, making water molecules in the skin to vibrate, which produces heat and causes discomfort. The closer a subject gets to the weapon the more pain is felt. This active denial technology developed jointly by the US Air Force Research Laboratory and the Department of Defense's Joint Non-Lethal Weapon Directorate (JNLWD) has had $40 million spent on it over a ten-year research period.[19]

- *Ship Defence Systems* – After the attack on the USS *Cole* in Aden, the US Navy's Force Protection Task Force was briefed by the JNLWP on relevant non-lethal technologies to enhance ship security. These included the Running Gear Entanglement System to help augment ship perimeter defence systems.

- *Unmanned Aerial Vehicles* – The development of technologies associated with UAVs that can carry mixed lethal and non-lethal payloads made rapid advances during the 1990s. For example, the Loitering Electronic Warfare Killer (LEWK) is designed to carry a variety of lethal and non-lethal payloads weighing up to 90kg.[20] Such a UAV would also be capable of suppressing of air defences ('aggressive' attack using bombs and 'passive' methods using jamming processes) and intelligence gathering. This confusion of roles would make it difficult for an enemy to know whether they faced a lethal or non-lethal threat and to frame their responses appropriately.

- *Civil Policing* – The use, and potential use, of NLWs by civil police forces world-wide received wider publicity during 2001, especially because of the anti-globalization rioting, their use in other civil disturbances (situations associated with sporting events[21] and ethnic clashes in cities) and a number of fatalities which occurred during more routine police operations.[22] Police in the UK and US are testing the A3P3 (Aerosol Arresting Agent/Pulse Projected Plume) gun which combines several non-lethal technologies in one weapons system – electric shock, pepper spray and video surveillance technology.[23] The weapon uses sensors to judge the distance of an attacker before releasing the 'correct' amount of pepper spray. If an attacker is electrically shocked at the same time, the resulting involuntary inhalation forces them to take in more of the pepper spray. If the user comes under personal attack, a switch on the gun can transfer the electric charge to pads on the user's protective clothing. The A3P3 is also fitted with tiny video cameras that can both record events and transmit 'real time' pictures back to the police control centre. Other developments include a high-powered glue gun which fires a pellet of compressed foam that expands on impact to about 30 times its original size, thus impeding the target's movement,[24] and a liquid stun gun where an electrical current travels along a jet of highly conductive water. This has a longer range than the usual Taser stunners (which just use electrical wires to deliver the charge), but is more cumbersome.

The Future of Non-Lethal Weapons

The majority of the articles contained in this book were first published in *Medicine, Conflict and Survival* [2001; 17 (3)]. They were developed from presentations at a conference in Edinburgh, December 2000, organized by the Bradford Non-Lethal Weapons Project,[25] which brought together a small working group of international experts reflecting a variety of perspectives and approaches under the heading of 'The Future of Non-

Lethal Weapons: Technologies, Operations, Ethics and Law'. Three extra pieces have been added to the collection – by Malcolm Dando (an original contribution), Victor Wallace [*Defense Studies* 2001; 1 (2)] and Brian Rappert [*Peace and Change* 2001; 26 (1)]. As the studies in this collection will show, whilst there are evident advantages linked with non-lethal weapons, there are also key areas of concern associated with the development and deployment of such weapons, including:

- Threats to existing weapons control treaties and conventions;
- Possible use of NLWs in human rights violations (such as torture);
- Possible harmful biomedical effects associated with NLWs;
- What some believe is a dangerous potential for use in social manipulation and social punishment within the context of a technology of political control.[26]

In the opening chapter, John Alexander challenges the very basis on which objections to non-lethal weapons are made, arguing particularly that most are based on emotions rather than facts. Whilst acknowledging that there are problematic areas, he argues that 'skeptics who find fault with non-lethal weapons almost invariably focus their attention on some relatively minute aspect of the problem or postulate abstruse theoretical implications' and that despite their faults these weapons do offer a real alternative to either doing nothing or killing people. For Alexander it seems surprising that the most strident objections to the implementation of non-lethal weapons have come from organizations that are ostensibly designed to protect non-combatants. He believes that these arguments are specious and, while technically and academically challenging, actually serve to foster an environment that will result in the deaths of many more innocent civilians. Opponents misconstrue technology with human intent. The reasons for use of force will not abate, therefore alternatives to bombs, missiles, tanks and artillery must be found. Non-lethal weapons are not a panacea, but do offer the best hope of minimizing casualties while allowing nations or alliances the means to use force in protection of national or regional interests. Alexander also introduces the concept of 'ergofusion' to support his argument that just because non-lethal weapons are available, it does not necessarily mean that they will lead to an 'increased propensity for conflict or more pain and suffering'.

Literature on 'non-lethal' weapons frequently contains assertions that more robust NLW development and use are needed because of the changing nature of military operations. These assertions are in conflict with international legal analysis of NLWs, which show international law restricting NLW development and use. In his chapter, David Fidler examines this conflict by (1) briefly analyzing the restrictive impact that international law has on NLWs, and (2) elaborating three perspectives about what the relationship between NLWs and international law should be. He outlines the moral foundations for existing international law on the use of force and armed conflict and sketches international law's current impact on NLW development and use. He then explores the compliance, selective change and radical change perspectives that emerge from discourse about international law and NLWs. The compliance perspective insists that NLWs comply with existing rules of international law. The selective change perspective seeks limited changes in international law to allow more robust use of NLWs. The radical change perspective sees in NLWs the potential to reform radically international law on the use of force and armed conflict. Identifying the three perspectives helps clarify future choices NLWs may present in international law and suggests that the future relationship between NLWs and international law will be more complex, controversial and dangerous than people may realize. Of importance here is the impact of what some refer to as 'future war' and the argument for radical change within international law to take into account the nature of new non-lethal weapons technology and the nature of the contemporary context within which they will be utilized.

Gerard Quille places the development of non-lethal weapons within the context of the revolution in military affairs (RMA) which has strategic, security, political and technological perspectives. Non-lethal weapons will be integrated into military strategies, especially those that have particular application in urban conflict environments. Of particular interest is the relationship between the deployment and development of large-scale weapons systems (the 'heavier and organizing' end of military technology) and non-lethal weapons within the framework of a second or radical phase of the RMA. Other developments associated with this include cyber-warfare, robotics, cyborgs and psychotechnology. The issue is introduced from a politico-strategic perspective, with a critique of the RMA debate as a starting point for discussion of the role of NLWs in conflict. Questions are set out for the politicians and strategists who are asked to formulate policy based on technology to be used in new political/social conflicts and the danger of neglecting other important dimensions of politics and strategy *vis-à-vis* present conflict is noted.

Whilst pointing out that there is now an extensive literature associated with NLWs and a more critical and analytic perception of their utility, Brian Rappert notes that there are still basic disagreements about the very nature of non-lethality. He examines and critiques four approaches or frameworks for evaluating the effects of existing and future non-lethal weapons derived from concerns over international security (military operations), human rights (Amnesty International), international humanitarian law (International Committee of the Red Cross – ICRC) and domestic policing (politics of technology). Given the hot debate and disagreement over terminology associated with NLW characteristics, Rappert proposes a typology 'affordances' of NLWs, characteristics that might serve as a basis for assessing technologies. By 'affordances' he means 'perceived properties of an artifact that suggest how it might be used' and 'any technology has a multiple number of perceived uses and effects'. So the notion of affordances can be used as a means of framing discussions about the effects of weapons. For Rappert this approach can 'usefully complement many existing, conventional analyses and act as a basis for commenting on the possible implications of yet unrealized technology'.

In his chapter, Steve Wright draws from two reports commissioned by the European Parliament,[27] which investigated 'sub-lethal' weapons as potential instruments of political control. Wright argues for a principled approach to set controls and regulations on the export and licensed production of 'non-lethal' crowd control technology which causes inhumane treatment, superfluous injury or unnecessary suffering. This is particularly relevant in the case of countries that are known to use them in violation of human rights. Wright also agues that non-lethal weapons are designed to 'appear' rather than 'be' safe and, since they augment rather than replace lethal technologies, their use can distort conflict and actually bridge the firewall between use of less-lethal and lethal technologies.

As Jorma Jussila notes in his chapter, society has entrusted the police with power and an obligation to enforce law, maintain order and protect law-abiding civilians. It is not always possible to refrain from the use of some physical force when dealing with criminal incidents. Selecting appropriate non-lethal weapons (or 'use of force instruments') to carry out these tasks presents a multi-faceted problem of balancing human considerations, judicial and societal requirements, with tactical needs and technological possibilities. It has also been found that the weapons themselves do not always live up to the manufacturers' claims. No matter what the incident is, a police officer is expected to protect the innocent, him/herself, colleagues and others in the vicinity and to cause no more harm than is justifiable and unavoidable. Whilst any weapon can be misused, most non-lethal weapons do have a legitimate use. Jussila argues that misuse, such as torture, is not a property inherent to the technology but an intentional behaviour of personnel equipped with weapons. Most police officers would want to use minimum force during their work, but also have the right to protect themselves. However, there needs to be more accountability for police use of non-lethal weapons and it is essential that thorough research and fair and credible controls on police weaponry are in place to deter misuse and to maintain the trust of the public.

'The troubles' in Northern Ireland have claimed the lives of 3,636 people over the last 32 years, including 302 police-officers and 644 soldiers. Of these deaths, 315 were attributed to the military and 52 to the police. Writing from the perspective of operational experience of non-lethal weapons in Northern Ireland, Colin Burrows provides an overview of the development of police strategies and tactics in policing violent situations, within the context of the characteristics of disorder and the type of weaponry (both lethal and non-lethal) used against the police force. The evolution of a new type of baton round (L21 plastic bullet) is described and the advantages of its use stated. However, it is also pointed out that very serious injuries will occur should it strike the head. Burrows stresses that in Northern Ireland the police are the lead agency and the military act in support of the police. The tests of what are 'appropriate', 'proportionate' and 'measured' responses to given situations are key to the public's perception of police action. Burrows concludes by arguing for a holistic approach for the police along a spectrum of responses from community policing to the more aggressive intervention sometimes required for dealing with situations of violence. As an alternative, the key features of Conflict Management are discussed and the importance of understanding the underlying causes of a conflict for law enforcement personnel is emphasized.

Jürgen Altmann questions claims made for the effectiveness of non-lethal technologies and argues for a more scientific analysis of effect of NLWs on their targets within a rigorous independent academic and public science setting. He claims that a systematic approach to preventive arms control is needed, which limits new militarily relevant non-lethal technologies before the corresponding weapons are deployed. That is, more attention should be paid to them at the stages of research, development and testing. A more transparent process is needed and independent scientists have a key role to play in providing the public with accurate information. This will require the establishment and funding of research institutions. To illustrate his argument that many of claims made about weapons are untrue, Altmann examines existing non-lethal technologies including chemical, biological, acoustic and kinetic impact weapons.

Nick Lewer and Tobias Feakin look at non-lethal weapons from the perspective of proliferation. Using India as a case study they show how, like any other weapon system, non-lethal weapons have followed an armament dynamic of supply-side and demand-side proliferation determined by both domestic and strategic issues. Picking up some of the issues raised in the chapter by Alexander, they ask why should we be concerned about the proliferation of NLWs and point to areas of serious concern. This is balanced by the legitimate reasoning and strong arguments put forward by Alexander and others for their wider use. Whilst it is inevitable that proliferation will occur, it is argued that more accountability and stricter controls are needed, especially with regard to states who are known for their civil liberties and human rights abuses.

Given the broad range of technologies which NLWs span and the multi-faceted, complex and inter-related implications for military policy and doctrine (especially within the context of peace support operations), Victor Wallace questions whether non-lethal weapons should be subject to specific arms control measures. Drawing from experiences in Somalia, Kosovo and other operational environments, NLW doctrine and the concept of Force Continuum have been developed to articulate their relationship with conflict intensity. This thinking has progressed to include the idea of a Conflict Intensity Continuum, to complement the Force Continuum concept. The Force and Conflict Intensity Continuums have become cornerstones of American NLW doctrine in the 'Joint Concept for Non-Lethal Weapons'. Wallace argues that the most realistic approach would be to examine NLW technologies in same manner as conventional weapons. He proposes the R^2IPE acronym (Reversibility of effect – 'Rheostat' capability – Information – Policy – Effect), as an additional assessment tool to help decide whether to develop and acquire a NLW system. He combines the R^2IPE elements with ICRC criteria which evolved from the Superfluous Injury or Unnecessary Suffering (SIrUS) Project. Wallace believes these criteria 'offer the opportunity to design out illegal features or simply to abandon weapon development before its effects become evident in the field. A window of opportunity exists that may ensure the mistakes of the past are not repeated in the future.' This combination of R^2IPE and ICRC could provide a more comprehensive framework within which to develop a method to appraise NLW systems without resort to specific arms control measures – an alternative to the 'difficult and tortuous process of evaluating the individual technology'.

Malcolm Dando begins by noting the concerns expressed in the late 1990s when it was realized that Iraq had developed large amounts of a chemical incapacitant (termed Agent 15) at the time of the 1991 Gulf War. A potential loophole in the Chemical Weapons Convention, that might be seen to allow the further development of such incapacitating agents for 'law enforcement' purposes, is then pointed out. A particular problem arises because such incapacitating non-lethal weapons appear increasingly attractive as armed forces become involved in many OOTW. Industrial countries had attempted to develop incapacitating chemicals such as BZ in the 1960s and 1970s, but lack of knowledge of the receptors affected in the central nervous system prevented success. The main part of the chapter shows how the advance of genomics, and its impact on neuroscience, has now opened up the route to successful development of effective incapacitants. The dangers of such developments are emphasized and the need to take preventive action, both inside and outside of the Chemical Weapons Convention, is stressed.

The common theme running through this volume is that there is a legitimate role for non-lethal weapons, both for civil and military applications. However, there is considerable disagreement as to the operational effectiveness of NLWs, the dangers such weapons pose to civil liberties, human rights and the threat to arms conventions and international law. As usual, a balance has to be achieved where the benign advantages of developing and deploying non-lethal weapons are not outweighed by their more malign effects.

Notes

1. *Policy for Non-Lethal Weapons*, United States Department of Defense, Directive No.3000.3, 9 July 1996.
2. Alexander J. *Future War: Non-Lethal Weapons in 21st Century Warfare*. New York: St Martin's Press, 1999; Dando M. *A New Form of Warfare: The Rise of Non-Lethal Weapons*. London: Brassey's, 1996; Dando M, ed. *Non-Lethal Weapons: Technological and Operational Prospects*. London: Janes, 2000; Lewer N, Schofield S. *Non-Lethal Weapons: A Fatal Attraction? Military Strategies and Technologies for 21st Century Conflict*. London: Zed Books, 1997; Morehouse D. *Nonlethal Weapons: War Without Death*. Westport: Praeger, 1996.
3. Lewer N. Benign Intervention and Non-Lethality: Wishful Thinking For the 21st Century. In: Dando M, ed. *Non-Lethal Weapons: Technological and Operational Prospects*. London: Janes, 2000.
4. See the chapter by Gerrard Quille in this book and also Metz S, Kievet J. *Strategy and the Revolution in Military Affairs: From Theory to Policy*. Carlisle: Strategic Studies Institute, US Army War College, June 1995; Metz S. *Armed Conflict in the 21st Century: The Information Revolution and Post-Modern Warfare*. Carlisle: Strategic Studies Institute, U.S. Army War College, 2000.
5. Freedman L. Britain and the Revolution in Military Affairs. *Defense Analysis* 1998; **14** (1): 55–66.
6. U.S. Joint Non-Lethal Weapons Directorate. Kosova-USA Use of NLW. *JNLWD Newsletter* 2000; 3rd Quarter: 1.
7. Military Shows Airlines Non-Lethal Protection Options: Riot Control Techniques Suggested for Air Crew. *Defense News* 28 Sept. 2001; see also discussions about a Robo-Lander anti-hijacking system, cockpit isolation systems and non-lethal weapons at http://www.iasa-intl.com/folders/RoboLander_files/non-lethalweapons2a.htm.
8. Kauchak M. Dazzled By the Light. *Armed Forces Journal International* July 2001. The Laser Dazzler is a green laser producing 'a light which envelopes, disorientates, confuses and temporarily blinds the subject'. It is claimed to be 'eye safe'.
9. Landmine Action. *Alternative Anti-Personnel Mines: The Next Generations*. London: Landmine Action, 2001.
10. Report of presentation by Dr Yvon Durant *et al.* at the Non-Lethal Technology and Research (NTAR) Symposium, Nov. 2000. See http://www.unh.edu/ntar/PDF/Durant2.pdf.
11. Pain S. Stench Warfare. *New Sci*, 7 July 2001: 42–5.
12. Albert S, Hitt W. *Intercultural Differences in Olfaction*. Battelle Memorial Institute: Remote Area Conflict Information Center, 2 May 1966. Part of DARPA's 'Project Agile'. Sunshine Project DOD Freedom of Information Request 01-F-1021. Quoted in Sunshine Report, July 2001. See also Pain S. Stench Warfare. *New Scientist*, 7 July 2001: 42–5.
13. The Sunshine Project. *Non-Lethal Weapons Research in the U.S.: Calmatives and Malodorants*. Backgrounder Series No.8, July 2001, http://www.sunshine-project.org.
14. See the chapter by Malcolm Dando in this book and also Dando M. *A New Form of Warfare: The Rise of Non-Lethal Weapons*. London: Brassey's, 1996; Dando M. *The New Biological Weapons: Threat, Proliferation and Control*. Colorado: Lynne Reinner, 2001.

15. *JNLWD Newsletter.* 2001; 2nd Quarter, http://www.jnlwd.usmc.mil/.
16. Arkin W. A Devil's Workshop? *Washingtonpost.com*, 10 Sept. 2001.
17. Kettle M. Revealed: The Weapon that doesn't Hurt Anyone. *Guardian*, 3 May 2001.
18. Brinkley M. The People Zapper. *Marine Corps Times.* 5 March 2001.
19. United States Air Force. *Active Denial Technology: Directed Energy Non-Lethal Demonstration.* Kirkland AFB, New Mexico: Air Force Research Laboratory, Office of Public Affairs, March 2001, http://www.de.afrl.af.mil/pa/factsheets.
20. U.S. To Test Combined Lethal/Non-Lethal UAV. *Jane's International Defence Review*, May 2001: 23.
21. For example, stingball grenades and rubber bullets were used in April during riots after a baseball match in Tucson, Arizona. The No.15 grenades make a loud noise and release .32 calibre rubber balls when they explode. See Ceja J. Non-Lethal Weapons Crucial to Breaking Up 4th Avenue Riots. *Arizona Daily Wildcat*, 19 April 2001.
22. Feakin T. *Research Report No.3.* Bradford: Bradford Non-Lethal Weapons Project, Aug. 2001, http://www.brad.ac.uk/acad/nlw.
23. Bright M. Riot Police to Get US Supergun. *Observer*, 3 June 2001: 23
24. Bamber D. Glue Guns Will Help Police Quell Violent Protestors. *Sunday Telegraph*, 3 June 2001: 10.
25. http://www.brad.ac.uk/acad/nlw/.
26. Ballantyne R. The Technology of Political Control. *Covert Affairs Quarterly*, 1998; **64**: 17–23; Ackroyd C *et al.* *The Technology of Political Control.* London: Penguin Books, 1977.
27. *An Appraisal of the Technologies of Political Control.* Luxembourg: European Parliament STOA Panel, Sept. 1998; *Crowd Control Technologies.* Luxembourg: European Parliament STOA Panel, June 2000.

An Overview of the Future of Non-Lethal Weapons

JOHN ALEXANDER

Finding fault with any topic is of course far easier than proposing solutions. The fundamental problem with the arguments against non-lethal weapons is that their proponents offer no viable alternative. Assuming that no utopian intervention will take place in the foreseeable future, humans will continue to engage in conflict, just as they have in the past and are today. Traditional weapons offer two difficult choices – do nothing or kill. Non-lethal weapons (NLWs) offer alternatives. In my view, those who oppose the use of non-lethal weapons by fiat must support either capitulation or killing. If they believe that humans will miraculously stop – or even lessen – their near ubiquitous aggression if non-lethal weapons simply did not exist, then we live on different planets and discussion is pointless.

In addressing the future of non-lethal weapons, I will cover both technical and philosophical aspects since they are completely intertwined. Unfortunately, those who enter into ethereal discussions have been accorded public sounding boards and raised alarms of the impending doom and obscuration of rights should NLW research move forward. The technologically developed world has a visceral attachment to dire circumstances and seems to react only when threats are posed. Reality is not a necessary ingredient – doom sells well. In addressing those academics who cast aspersions, I would quote my good friend, the late Ben Rich, former president of the famous Lockheed Skunk Works, when he said: 'What have they ever built?' As you will see, my version of Rich's question is, 'Compared to what?'

Since we are addressing 'current thoughts' this argument is time dependent and may change as the global situation changes. In the past decade non-lethal weapons have moved from virtual obscurity to a position of some prominence and controversy. There are those who see the situation differently and I believe rational debate will be both lively and healthy.

Critical Issues

I feel obliged to address certain issues just to get them out of the way. These issues arise in every discussion in which new participants join in the non-lethal weapons debate. They have been examined in great detail.

Non-lethal weapons are a goal based on intention. There are no perfect weapons that will never kill anyone under any circumstance. The goal – not an absolute – is no *unintentional* loss of life. Unintentional is important because there will be situations in which we intend to kill some people but not others; that is the way of war. There is also a goal for controlling the level of physical damage inflicted. At times the level of acceptable damage may be very high. The intent is control of damage. In many of the conflicts in which we have engaged recently, it is the victor who pays the vanquished in re-establishing their normal life. Certain items in the infrastructure are known to require long lead times to replace. In general it is better to damage selected items that can be replaced in shorter periods of time so that vital services can be restored expeditiously upon the end of hostilities and thus minimize the inconvenience to the civilian population. We also want to expand options available to commanders although not everyone agrees that this is a good thing. There are those who would have the options be kill or be killed and nothing less. Some of the other previously discussed issues include:

- *Semantics* – Non-lethal is not a perfect word nor are non-lethal weapons perfect. Many other words have been tried. Generally we come back to non-lethal. It serves the purpose of describing the intent of the weapons. Some people will die if such weapons are used improperly. That point is acknowledged. However, semantic debates simply detract from the core issues.

- *Premature Use of Force* – Some people suggest that having non-lethal weapons may cause earlier armed intervention before other diplomatic means have been exhausted. This is a policy issue, not a weapons issue. However, force should never be used unless clear objectives are established. There is little credence in the notion that non-lethal weapons will make mature democratic regimes more adventuresome. If they do, the problem is with the election of their politicians, not the non-lethal weapons.

- *Cost-Effectiveness* – In a time of reduced budgets, some argue we should put the money into more lethal weapons but there are times at which having non-lethal options are essential and thus cost effective. They may mean the difference between containment of the situation and escalation into wider conflict, the expense of which would be much greater. Budget tradeoffs are always tough decisions and non-lethal weapons must be able to prove their worth when compared against other systems. As experience is gained, quantitative cost-benefit analysis will improve.

- *Own Vulnerabilities* – Some of these systems can be used against friendly infrastructure. However, the vulnerabilities are present with or without non-lethal weapons. In *Future War*[1] I introduced the concept of cerebralcentrism. It refers to a predominantly – but not exclusively – American trait in which we believe ourselves to be the only smart people in the world. The argument goes that if we don't develop these weapons, no one else can. Vulnerabilities are inherent in democracies and they will be exploited. Countermeasures should be developed simultaneously to any new weapons system. Physics and chemistry work the same for everybody and there are very few technical secrets that last very long.

- *What If They Shoot Real Bullets* – By doctrine, non-lethal weapons are never employed without lethal support. Soldiers should not be put in harm's way without adequate protection. If an adversary shoots real bullets then troops will immediately respond in kind. Non-lethal weapons provide a warning not available if only lethal weapons are present. Conversely, we have made it clear that availability of non-lethal weapons does not infer they must be used before lethal weapons are employed.

Why Do We Need Non-Lethal Weapons?

There are some strong arguments for the development and acquisition of non-lethal weapons.

World Geopolitical Situation

The world has changed dramatically during the past decade. At the moment we no longer face the threats of mutual annihilation or national survival. However, small wars and internal conflicts are constantly erupting in many areas around the globe. In addition, non-governmental organizations (NGOs) are playing increasingly important roles to the point that they can influence regional stability. Conflicts of the future will not necessarily be with other nation-states.

Maturity of Technology

Many of the non-lethal weapons now being developed were postulated several decades ago. The technologies of that day were not sufficiently advanced to accomplish the task. With new materials, dense energy storage and, most importantly, precision delivery mechanisms, we can employ new, effective non-lethal weapons.

Operational Experience

Non-lethal weapons concepts moved forward when commanders in the field experienced peace support operations. They returned from engagements and demanded requirements were written for the development of non-lethal weapons. They wanted options between the extremes of doing nothing or killing people. The more commanders became frustrated because of the operational limitations placed on them by strict rules of engagement, the more they demanded that effective non-lethal weapons be made available.

Operation 'Desert Storm' Technology Demonstration

There was a downside to 'Desert Storm'. Our technology proved so overwhelming that potential adversaries around the world took note. No reasonable enemy is going to allow an extensive coalition to be developed, logistical infrastructures put in place, overwhelming firepower to be massed and then permit us to set the time and place of the engagement. Traditional threats will remain and adversaries will attempt to fight on their terms, not ours. Whenever possible those adversaries are going to co-locate with disinterested noncombatants and make the use of our overwhelming firepower morally difficult to justify. The notion of asymmetric warfare has become widespread.

Non-lethal weapons and concepts cannot be viewed in an isolated environment. There are still conventional threats that demand a highly mobile and extremely lethal response. These include nation-states such as Iran, Iraq and North Korea, to name some of the obvious ones. Blunt aggression must be met with the force necessary to overcome it.

In addition, there are both state-sponsored terrorism and terrorist organizations that cannot be controlled by their host nation. It is these activities that are embedded in innocent civilian populations that are most difficult. We know how to deal with these threats once they exceed our tolerance level. The response will entail collateral casualties, no matter how unfortunate that might be. The situation will become exacerbated by population increases and shortages of potable water. Information is critical and there needs to be greater emphasis placed on intelligence about terrorist's plans and activities. Non-conventional threats are endemic. These include transnational organizations, some of which have the power of nation-states. For instance, organized crime is estimated at a worldwide annual income of one trillion dollars. Of that $300–400bn is in illegal drugs. That level of funding is sufficient to destabilize regions. In addition there has been the near ubiquitous devolution of former countries leading to mass migration, ethnic and religious conflicts and other sub-national strife. As demonstrated in the Balkans, central Africa, southeast and southern Asia, these situations are very difficult to handle.

The non-lethal weapons systems that have been developed under the auspices of the US Joint Non-Lethal Weapons Directorate (JNLWD)[2] have been designed to assist in peace support operations. There were two logical reasons for their conception. First, as troops were being deployed in peace support operations they were in dire need of these weapons. Second, many of them were fairly easy to develop and field. Prototypes already existed and only minor modifications were necessary before they could be placed in the hands of troops. Also, there was considerable pressure to get something fielded to meet the rising expectations of troops and commanders.

Strategic Implications of Non-Lethal Weapons

From the inception of non-lethal weapons concepts, some of us believed that they would play a role across the entire spectrum of conflict. In the past two years some senior military leaders have started to voice that opinion. Here is a specific example of why we need a strategic non-lethal weapon's capability. For sometime prior to 'Desert Shield'/'Desert Storm', Saddam Hussein had his eye on Kuwait. Even after occupying Kuwait and the coalition had begun amassing forces, Saddam did not believe we were serious about attacking his forces. Basing his notion on American social politics he did not believe that we had the *will* to fight. Obviously he was wrong.

How can non-lethal weapons play a strategic role in similar situations? No one doubts the American/NATO military capability. The questions that remain are whether or not there is both the *intent* and the *will* to use force in a given situation. Non-lethal weapons designed to attack an adversary's infrastructure can provide that message. For example if the communications or transportation systems are attacked, it would indicate the intent to use the force necessary to achieve our objectives and that we have demonstrated our will to do so. Of course we are cognizant of the potential for secondary and tertiary casualties. These must be factored into the decision of whether or not to use force at any level.

The wise adversary would then capitulate and acquiesce to the stated demands, unless of course they were expecting to escalate to war all along. In the event of war, the non-lethal attacks against the infrastructure would have begun the weakening process. From a political perspective, it must be assumed that if non-lethal weapons do not achieve the desired goals, or the adversary immediately chooses a shooting war, then the necessary forces and the war plans are in place to instantly respond. Decisions about such use of force should not be made incrementally – in other words, no bluffing.

In March 2000, the National Defense Industrial Association (NDIA) held Non-Lethal Defense Conference IV (NLD IV). There were significant differences between this and the earlier conferences. In the prior sessions, most of the weapons developers had been small companies working in niche markets but this time major defence contractors took an active part.

This change signalled a significant shift in thinking. Funding has always been a key issue. As with any startup programme it takes time to raise the necessary financial support – in the US Defense budgets are projected as much as seven years in advance. When a totally new programme is initiated it is very difficult to obtain adequate amounts of money. The new programme must compete against existing programmes and demonstrate why the money would be better spent in the new area rather than on the established programme. It is a zero-sum game and the infighting can be vicious. For the new programme to be funded, some other contractor must lose that funding.

Further, the process is entirely *requirements* driven. This means new technical developments are funded only if they meet a formally established need that has been articulated in a requirements document. With tight funding it is imprudent to develop a technology just because it seems like a good idea. Defence contractors are generally economically motivated. During the formative years they recognized that the relatively small amount of research and development funding was not backed by substantial procurement money which is where industrial profit is made. Therefore, while some companies did choose to follow the information available pertaining to non-lethal weapons, they did not make any corporate commitments to participate.

At NLD IV Lockheed-Martin, Jaycor,[3] and PriMEX[4] all provided presentations on various systems and concepts. Recently, Boeing has been touting a non-lethal version of the Advanced Tactical Laser[5] – this is a major end item that comes with a big-ticket price tag and would only be developed if there were requirements for use in many different scenarios.

Operational Requirements

There is one practical question raised by troops confronted with angry civilians. That is, 'Is this technology better than a rock?' Specifically, they want weapons that provide sufficient standoff so that they can out-range a rock thrown by a strong teenager (about 180 feet, not counting bouncing). For many of the simple kinetic and shock technologies the answer has been no although ranges are improving. Therefore one of the key requirements generated by troops was to be able to employ non-lethal weapons from distances outside rock-throwing range.

One advance that provides the appropriate standoff comes from paintball war-games. To provide more realistic games, developers have been testing compressed gas guns with extended ranges. They are now fairly effective up to ranges of 100 meters. While there is some use for marking dyes, paintballs are not of great interest. However, micro-encapsulation technology is advancing and there are a variety of agents that can be projected with adequate accuracy to hit the designated targets without seriously affecting those around them. While the compressed gas launchers are capable of obtaining the desired distances, work is currently being done on improving flight stability of munitions. This stability will allow sufficient accuracy for the soldier to hit designated individuals rather than employing the launcher as an area weapon.

Changes are taking place in the development of electrical shock weapons that are designed to incapacitate selected individuals. Air Taser[6] has an improved system that provides sufficient power to stop an individual, even if they have been made pain-resistant on drugs. Still, the increased range of 21ft does not meet the rock-throwing test. There are some advances on the horizon that will probably overcome this problem.

Militarily, there is no unanimity concerning the need or efficacy of non-lethal weapons. Some military officers and politicians argue that peacekeeping is not an appropriate mission for the military and that only extremely lethal weapons are necessary. This is a moot point. The US and foreign militaries have been involved in peace support operations. They are currently engaged in such missions in Bosnia, Macedonia, the Sinai and elsewhere. They will continue to be committed to peace support operations. It is therefore imperative that the soldiers be provided with tools appropriate for such a complex and sensitive job.

Taking on the Critics

Non-lethal weapons are not a panacea. There are legitimate concerns about their development and use. These include the likelihood of producing unintended death or serious injury and inappropriate use by untrained personnel. However, there have been other vociferous complaints that are groundless. For instance, one concern about rubber or wooden bullets has been that they inflict pain, can cause bruising and, in rare instances, result in death – all true. However, the issue that is not addressed by the opponents of non-lethal weapons is 'Compared to what?' Non-lethal weapons are meant to be adjuncts to lethal weapons. They are not to be used without provocation or proper authority. Therefore, if provocation exists and non-lethal weapons are not available, the perpetrator is likely to be shot with a lethal weapon.

Another negative application cited by critics is use of non-lethal weapons as implements of torture. This complaint is especially prevalent for electrical stun guns. In reality, almost any item can be used to inflict pain on another human being. Non-lethal weapons may slightly increase this potential but the problem can be handled with adequate training and supervision. The fundamental issue is training and supervision, not technology. There is nothing about non-lethal weapons that makes them inherently more dangerous than any other implement.

Finally, there is a small but vocal group of conspiracy theorists that view non-lethal weapons as tools for illegally controlling the civilian populace. This argument also fails the test of logic. Sufficient force already exists to control large segments of society. It is the moral fibre of our people that keeps us free, not technology.

There are issues that have been introduced in discussions about non-lethal weapons that are totally specious. The concept of *benign intervention* is one of those. No responsible military official has ever suggested that forces should be introduced into conflict in anything that vaguely emulates benign intervention. If military forces engage in conflict, people are going to die. The question is, how many? The notions of a 'bloodless war' or 'a war without death' are utter nonsense. The issue of benign intervention offers another example of factually ungrounded emotional discourse that detracts from the serious business at hand. It might be amusing if people weren't dying while this debate continues. The principal difference in thought about the use of non-lethal weapons revolves around responsibility and accountability. In general, it is those people with operational responsibilities that recommend use of non-lethal weapons. The detractors, while often altruistically motivated, rarely have any responsibility for implementing their words or the consequences thereof.

Let me cite a specific example and the recent turnaround in thinking. Steven Aftergood wrote a scathing article titled 'The Soft-Kill Fallacy' decrying non-lethal weapons. The large print sidebar stated: 'The idea of "non-lethal weapons" is politically attractive and purposively misleading.' He went on to say:

> The futuristic aura of many non-lethal weapons is seductive, and their advent has been heralded uncritically by many media reports of kinder, gentler weapons. But basic political, legal, and strategic questions about the utility of the non-lethal thrust remain unanswered – sometimes even unasked.[7]

That comment is as false today as it was then. The fact of the matter was that from the beginning many of the moral and legal issues were being raised, not by reporters, but by weapons developers and military officials with responsibility for their actions. Competent people in positions of authority fiercely debated these issues from the inception of the concepts. Fostered by Hollywood and supported by the media, it is *nouveau chic* to believe that military officials are not competent to think critically about complex issues. In reality, nothing could be further from the truth.

In an attempt to denigrate further the field of non-lethal weapons, Aftergood resorted to the time-honored media approach – an *ad hominem* attack. After ten lengthy paragraphs (more than a third of the article) about my background he concluded with:

> John Alexander is by all accounts a resourceful and imaginative individual. He would make a splendid character in a science fiction novel. But he probably shouldn't be spending taxpayer's money without adult supervision.

On 2 November 2000, Steven Aftergood sent me an email in which he bridged the mental gap between philosophy and reality. He wrote:

> I was going to say that I have thought of you a couple of times over the past weeks. In particular, I felt that I may have done a serious disservice in my criticism of non-lethal weapons. Watching the violence between the Israelis and Palestinians unfold, with the accompanying deaths, I couldn't help but wish for the easy availability of non-lethal weapons. The largely hypothetical concerns I had suddenly seemed less compelling. Maybe it is not too late to do something about it.

The Road to Hell is Not Paved with Non-Lethal Weapons

I have been continually amazed at the organizations that have become the primary opponents to non-lethal weapons. Surprisingly, they are usually the organizations associated with offering assistance to the beleaguered but their objections bring them into intellectual alliance with criminals and terrorist groups.

The International Committee of the Red Cross (ICRC) seems to be leading the international charge. Through their policies attempting to halt development and use of NLWs, the ICRC, by default, comes down firmly in support of killing people. Their concerns about possible subjugation and unnecessary suffering are both misguided and counterproductive. Another organisation, vocal in their opposition, is the American Civil Liberties Union (ACLU). Similar to the ICRC, the ACLU employs erroneous and emotional arguments to support their claims. In so doing they too, *de facto*, support killing as the only viable alternative in use of force. In many of the cases they cite they seem to prefer that no action be taken. In general, civilian populations are fed up with criminals. Doing nothing is not a viable option.

Terrorist groups, while not openly vocal on the topic, are well aware of the threat to their operations. Non-lethal weapons offer the ability to strike at selective targets and minimize casualties.

Contrary to protestations, intent is everything. At the Jane's Non-Lethal Weapons conference of 1999, Robin Coupland, an ICRC representative, claimed the 'the road to Hell is paved with ...'[8] Meaning that non-lethal weapons, however well intended, are leading toward painful consequences. This thinking, however well intended, is misguided. Too frequently there is an attempt to blame technology (NLWs, mines, etc.) for problems that are based on the conduct of people. Contrary to the ICRC assertion, *intent* is omnipresent and is the most important factor to be considered. Barring an end to voluntary violence (an objective unlikely to be achieved while humans are involved) the best that can be done is to control the use of force.

My recurrent theme is 'compared to what?' Since violence will continue, what are the alternatives? Without non-lethal weapons we are left only with lethal ones.

Ergofusion

Speaking neo-linguistically, I would like to introduce the concept of ergofusion. It is defined as 'the misidentification of casual relationships.' Ergofusion is very prevalent in the field of non-lethal weapons where it is assumed that the availability of these systems will necessarily lead to increased propensity for conflict or more pain and suffering. A classic example of ergofusion can be found in a popular bumper sticker that reads: 'Guns don't kill people, people kill people'. Guns, like all material technologies, are inert objects that do nothing without the direction of an operator. Guns may be used for many purposes – some good, some bad. The purpose or intent is provided by the operator and is not found in the weapon.

Another extreme argument is that the use of gas, such as a riot control agent, against citizens is often equated by the media to the use of lethal gas in the Nazi death camps. Such an article was published in a Southern California paper when it was announced that the private security guards would be armed with pepper spray. The emotion laden on such chemical agents is both enormous and unfounded. The California ACLU has made extraordinary claims about the number of deaths attributed to the use of oleoresin capsicum (OC). The facts are that most of the deaths occurred from other causes – usually positional asphyxiation – and OC was not even a contributing factor.

Technologies do not cause bad behaviour. It is people who use technologies for evil purposes that exhibit bad behaviour. Blaming non-lethal weapons for increased conflict or human suffering is the quintessential example of ergofusion.

A Reconsideration of Issues

The arguments that highlight concerns about non-lethal weapons can be summarized as follows:

- What problem are you solving? This is non-trivial. In my view, the arguments against most non-lethal systems become isolated and focus on specific aspects of a given technology. The bigger picture of the desired outcomes for the current conflict is lost.

- Most arguments against non-lethal weapons are based on emotion versus facts.

- We blame technology for human problems. Most abuse of non-lethal weapons stem from inadequate training and supervision. I personally advocate a ban on the export of the most commonly used instrument of torture but the US Tobacco Lobby is too strong. There is nothing that cannot be misused.

- Chemical and biological agents have peaceful purposes. Despite the ill-advised arguments about CW/BW treaty violations, antimaterial agents will be legally developed. In fact, biochemistry for bio-remediation is growing at a near exponential rate. This is quite legal and will continue.

- Many future adversaries will not be signatories to treaties. This means that they will not feel bound by their provisions. It has already been demonstrated that many countries cannot, or will not, control elements resident within their geographic boundaries. Treaties may make politicians feel good but will be virtually meaningless. If asked which treaties have been broken, the correct answer is all of them.

- If we have the wrong treaties – change them. That statement alone can evoke wrath on an unparalleled scale. I have noted that Europeans, more so than Americans, seem to cling tenaciously to the concept that laws and treaties assume the power of a preordained edict of God. While well intended when designed, there are treaties that have outlived their usefulness. While they may keep lawyers gainfully employed, the downside is that real people may be made to suffer because of them – the exact opposite of the intention of those drafting those treaties. Even worse, treaties may provide an unwarranted degree of psychological security when the probability of infraction is actually very high.

Issues about Non-Lethal Weapons that Need to be Addressed

There are a number of important subjects that should be addressed. In general they involve effectiveness, casualty acceptability limits and rules of engagement.

Effectiveness

This is one of the most important issues facing commanders. They need to have high assurance that non-lethal weapons will perform to accepted standards. There are several questions of the effect of various types of non-lethal weapons on humans which include the wide range of variation that is likely to be found in a diverse group of civilians. We need to ask questions such as: What are the psychological effects of having non-lethal weapons present? Do non-lethal weapons serve as a deterrent to further aggression and, if so, under what circumstances? Will any non-lethal weapon have an adverse impact on the environment?

Casualty Acceptability

Unlike law enforcement, military operations are likely to have some level of casualties that are deemed acceptable by the politicians who commit troops in these situations. It is highly likely that the level of acceptable casualties will be based on such politically incorrect factors as race or religion. For example, in Somalia there were thousands of collateral casualties while the US was involved; can you imagine allowing that number of casualties in Western Europe? Another related factor may be the visibility of casualties. If CNN and other news organizations are filming these deaths, the acceptance level is likely to decrease. So what are the real factors that will determine the level of collateral casualty acceptability?

Rules of Engagement

There is a need to determine when non-lethal weapons should be used and when troops in the field should be provided with the rules of engagement. How are those rules of engagement implemented? Who has the authority to establish the rules of engagement and under what circumstances, and by whom, may they be amended? Under what circumstances should troops make the transition from non-lethal to lethal weapons? Is it possible to retreat from lethal combat to non-lethal weapons and under what circumstances should non-lethal weapons options be bypassed and the initial combat be engaged with lethal force?

The History Channel

A new issue has been emerging over the past few years. I call it 'The History Channel Effect'. Of course, it is well established that the history of conflicts is written by the winner. We are currently witnessing cathartic urges to produce revisionist versions of prior events. The propensity for invoking The History Channel Effect will further complicate future conflicts as commanders dutifully record every aspect of the actions to cover their rear. The CNN Effect is well known and there is often a cameraman with every squad who reports to the world in real-time. There is also an adverse impact of revisionist history whose attributes include:

- Poor/selective documentation by reporters – Rather than spending time researching, reporters either use the reports provided or select the facts that support their position.
- High on emotional impact – They are selling a product and emotion brings in higher ratings.
- Lack of combat experience by reporters – Few reporters have actually engaged in combat but they all think they have the innate sense to provide the truth.
- Revised standards of conduct – We have repeatedly experienced re-evaluation of historical events based on today's standards.

Several examples of media reports serve to illustrate my History Channel Effect point:

- Tail Wind – Based on a CNN report that claimed we targeted US servicemen and employed lethal gas during Vietnam. The incidents didn't happen.
- The Battle of Rumaili – Based on erroneous reporting of events from 'Desert Storm' in which senior commanders were personally castigated for having violated the rules of war.
- The Hurricane – A popular movie based on falsehoods in which a convicted killer had his sentence overturned. The facts indicate that Rubin Carter probably was responsible for the murders.

In Summary

There are discernable trends in the development and application of non-lethal weapons. The most important is that troops being deployed see the need for them. This is driving the formalization of the requirements process. Next, there is a broader vision of applications at tactical, operational and strategic levels. The broader vision and potential for increased funding are attracting larger defence contractors to become more actively involved. At the tactical level there are advances being made for weapons that provide minimum standoff.

At the same time, concerns about the efficacy of non-lethal weapons continue to be raised. These are, in my view, misguided and based on abstract philosophy rather than real fact. Any item may be used for torture and non-lethal weapons are no better or worse than the people who use them. As shown by 50 years of Soviet domination over Eastern Europe, sufficient lethal force is available to subjugate large numbers of people. Non-lethal weapons add little to that equation. It would be a serious mistake to allow this philosophical debate to dominate the issues surrounding the development of non-lethal weapons. Unfortunately this could happen. However, there are issues that should be discussed. They include effectiveness, casualty acceptability limits and rules of engagement. In a world in which mercurial adversaries are increasingly difficult to identify and locate as they commingle with innocent civilians, alternative weapons will be required. They will not be perfect but they can reduce the overall loss of life. From peace support operations through strategic applications, non-lethal weapons will be a part of that future.

Notes

1. Alexander J. *Future War: Non-Lethal Weapons In 21st Century Warfare*. New York: St Martin's Press, 1999.
2. Joint Non-Lethal Weapons Directorate. http://www.marcorsyscom.usmc.mil/jnlwd.
3. JAYCOR. http://www.jaycor.com/.
4. PRIMEX. http://www.army.technology.com.
5. Boeing Tactical Laser. http://www.boeing.com/news/releases/1999/.
6. Air Taser. http://www.airtaser.com.
7. Aftergood S. The Soft-Kill Fallacy. *Bull Atom Sci* 1994; 50 (5): 44–5.
8. Coupland R. NLWs: Some Medical, Tactical and Legal Concerns. Presentation to Jane's International Conference, Non-Lethal Weapons: Fielding NLWs in the New Millennium, London, 1–2 Nov. 1999.

'Non-Lethal' Weapons and International Law: Three Perspectives on the Future

DAVID P. FIDLER

The first perspective, the compliance perspective, holds that the tension between the need for non-lethal weapons (NLWs) and international law is resolved in favour of international law because NLW development and use should meet all existing legal requirements. The second perspective, the selective change perspective, posits that limited changes in international law may be required to allow NLWs to be used where they are needed for military and humanitarian purposes. The third perspective has not, to my knowledge, been advocated or discussed in NLW literature, but it emerges from discourse about NLWs. The third perspective senses the potential and the need for radical changes to international law in NLW development and on the use of force and armed conflict. This position I call the radical change perspective.

The three perspectives form a continuum illustrating potential impacts of NLWs on international law: from little or no impact (compliance perspective) to significant, near revolutionary impact (radical change perspective) (see Figure 1).

FIGURE 1
IMPACT OF NLWs ON INTERNATIONAL LAW

Compliance Perspective	Selective Change Perspective	Radical Change Perspective

NLW development and use has not reached a stage where conclusions can be reached about what kind of impact these weapons will have on international law. However, the analysis of the three perspectives is useful because it focuses attention on an important issue that NLW development and use will face – the impact of international law. In addition, the three perspectives help clarify moral and legal choices that may appear in the future if NLW technologies continue to develop and find favour within military and political circles.

International Law on NLW Development and Use[1-5]

The Moral Foundation of Existing International Law

Interest in NLWs has arisen predominantly in response to what military experts refer to as the changing nature of warfare and military operations. Weapons and tactics designed to fight past wars do not provide adequate options for military forces struggling to adapt to more complex missions, such as peacekeeping and 'military operations other than war'. This dynamic, changing military environment contrasts with the historical continuity and stability of the moral and international legal rules on the use of force and armed conflict. While international law in these areas has changed over the centuries, there is a remarkable consistency between age-old moral principles and the modern rules of international law.

The moral foundation for contemporary international law on war is the Christian Just War Theory (CJW theory). The theory placed restraints on engagement in warfare and the use of force in armed conflict.[6] In terms of engaging in warfare, the CJW theory posited that a leader could only resort to war for a just cause, such as the re-establishment of peace or to redress a serious injury to one's country.[6] War could not be the province of personal revenge or ambitions of national or personal aggrandizement.

If a leader justly resorted to war, then the CJW theory regulated how force could be used. First, militaries could not use force intentionally against civilians. Second, the force used against legitimate military targets had to be proportional to the threat faced and the stakes of the conflict. Third, when a use of force against a military target is likely to produce evil effects, such as the death of civilians, such use is still acceptable if the evil consequences are unintended and if the legitimate objectives of the use of force outweigh the unintended evil effects.[6,7]

Contemporary international law on the use of force and armed conflict embodies the CJW theory's tenets. Table 1 shows the connection between the CJW theory and international law.

NLWs and International Law on the Use of Force by States

NLW literature discusses the possibility that development and deployment of NLWs may lead to states using force more frequently, thus bringing international legal rules on the use of force under increased pressure.[4] The prohibition on the use of force except in cases of self-defence and United Nations Security Council military actions[8] proscribes using 'non-lethal' force in other contexts, such as humanitarian intervention, anticipatory self-defence and attacks on terrorist organizations.

TABLE 1
RELATIONSHIP BETWEEN CHRISTIAN JUST WAR THEORY AND
MODERN INTERNATIONAL LAW

Tenet of the CJW Theory	Manifestation in International Law
A leader may only engage in war for a just cause	A state is prohibited from using force in international relations, except in cases of individual self-defense, collective self-defense, or under authorization from the UN Security Council in response to a breach of international peace and security.
Military force may not be used intentionally against civilians.	Military forces are prohibited from attacking civilians and civilian targets.
The use of force against legitimate military targets must be proportional to the threat and the stakes of the conflict.	1. Legitimate uses of force, such as in self-defense, must be proportional. 2. Militaries may not use weapons that cause superfluous injury or unnecessary suffering to enemy combatants. 3. Military forces may not attack combatants who are incapacitated, disarmed, and who pose no military threat (*hors de combat*). 4. Military forces may not use weapons that produce widespread, long-lasting, or severe impact on the environment.
The unintentional killing of civilians during armed conflict is acceptable as long as the legitimate ends of the use of force outweigh the unintentional loss of life.	Civilian collateral damage from attacks on legitimate military targets is not illegal as long as the damage was not intentional and efforts were made to avoid such collateral damage.

International Law and the Development of NLWs

International arms control treaties restrict the development of certain NLWs. The Geneva Protocol and the Biological Weapons Conventional (BWC) prohibit the development, production and use of any biological weapon, whether lethal or 'non-lethal'.[9, 10] A protocol to the Inhumane Weapons Convention prohibits the development and use of blinding laser weapons.[11] The Land Mine Convention prohibits the development and use of anti-personnel landmines that kill, injure or incapacitate.[12]

The impact of the Conventional Weapons Convention (CWC) on the development of 'non-lethal' chemical weapons is more complex. The CWC prohibits the use of 'non-lethal' chemical weapons against enemy soldiers.[4, 13] It allows the use of anti-material 'non-lethal' chemical weapons so long as such weapons do not cause permanent harm to or incapacitate humans.[4, 13] The Convention prohibits using 'non-lethal' riot control agents as a method of warfare.[13] The CWC allows chemical weapons such as tear gas to be used for 'law enforcement purposes'[13] which are not defined in the treaty. Controversy exists as to whether military forces can legitimately use chemical weapons in peacekeeping activities that are akin to 'law enforcement purposes'.[4] In sum, 'the overall impact of the CWC on 'non-lethal' chemical weapons is significant'.[4]

Laws of War and NLW Use

• Prohibition on Attacking Civilians – International humanitarian law (IHL) prohibits military forces from attacking civilians.[14] Weapons must be targeted only at military objectives. This prohibition makes no distinction between lethal and 'non-lethal' uses of force and thus applies with equal force to NLW use against civilians.

• Prohibitions on Attacking Combatants Who Are *Hors de Combat* – IHL allows NLWs to be used against combatants but it prohibits military forces from attacking combatants who have been incapacitated and no longer pose a military threat.[15] NLWs would create incapacitated enemy combatants who could not be attacked. Military forces would have to devote significant personnel and resources to cope with casualties from NLWs[16] in order to comply with the *hors de combat* principle and prevent the incapacitated soldiers from regaining the ability to be a military threat. In connection with NLWs, the *hors de combat* principle imposes on military forces negative duties (no attacks on incapacitated combatants) and positive duties (addressing the needs of the incapacitated).

• Prohibitions on Weapons that Cause Superfluous Injury or Unnecessary Suffering – Military forces cannot use weapons that cause superfluous injury or unnecessary suffering in enemy combatants.[14, 17-19] This principle is the basis for the bans on blinding laser weapons and anti-personnel landmines. Other NLWs that may be developed, such as electromagnetic, microwave and acoustic weapons, may not be deployed if they violate the superfluous injury or unnecessary suffering principle.[4] At present the health effects of many NLW technologies are not completely understood;[20] therefore, this rule of international law holds an important position in future NLW development.

• Prohibition on Weapons Causing Environmental Modification – International law contains rules that prohibit the use of weapons that result in widespread, long-lasting or severe effect to the natural environment.[14, 21] NLWs, such as defoliants, soil de-stabilizers and weather modification techniques, will be prohibited by international law if they produce widespread, severe or long-lasting effects on the natural environment.

Summary of International Law's Impact on NLW Development and Use

The above sections outlined the restrictions that international law place on NLW development and use. It has been argued that, under current international law, there is no good reason to think in terms of lethal versus 'non-lethal' weapons because international legal principles must be applied to all weapons.[4] The existing rules of international law, therefore, pose hurdles to NLW development and use.

To Comply or to Change?
Tension Concerning International Law in NLW Literature

One feature of NLW discourse on international law is the tension between the compliance perspective and the selective change perspective. This section explores these two perspectives and their implications for international law's role in the development and use of NLWs.

The Compliance Perspective

The compliance perspective holds that NLW development and use should satisfy the relevant rules of existing international law. This perspective appears in official NLW policy documents. For example, NATO's NLW policy states that:

> The research and development, procurement and employment of Non-Lethal Weapons shall always remain consistent with applicable treaties, conventions and international law, particularly the Law of Armed Conflict as well as national law and approved Rules of Engagement.[22]

Similarly, the United States Department of Defense NLW Directive provides that all NLWs shall undergo a legal review to: 'ensure consistency with the obligations assumed by the U.S. government under all applicable treaties, with customary international law, and, in particular, the laws of war'.[23] Part of the compliance perspective is that many existing NLWs do not run afoul of current international legal rules. Alexander argues, for example, that 'in general, most non-lethal weapons meet all these [international legal] tests'.[1] However, evidence already exists to illustrate that conflicts between NLWs and international legal rules can and will occur. Coppernoll reported that the US Navy Judge Advocate General did not approve biologically-based NLWs because the microbe 'category of weapons was held to violate the Biological Weapons Convention'.[2]

Caveats about the compliance perspective are necessary. First, governments and international organizations have little choice but to proclaim that their NLW policies will conform to international law, but such proclamations do not mean that governments, alliances and their respective military forces are happy about the restrictive impact of international law on NLW development and use. Second, as indicated above, the impact of some international legal rules on NLW development is ambiguous because of divergent interpretations of treaties, particularly the CWC. To proclaim compliance may mean compliance with one's controversial interpretation of a relevant rule of international law. Third, the compliance perspective does not prevent governments from attempting to change international legal rules to be more accommodating to NLW development and use. If the rules happen to change, then the compliance perspective remains conveniently valid.

In sum, the compliance perspective places obedience to existing rules of international law above NLW development and use, even if this prevents certain NLWs from development or deployment. Therefore the compliance perspective does not accept that the changing nature of warfare and war-fighting technologies undermines the existing moral and legal principles regulating the use of force and armed conflict.

The Selective Change Perspective

The selective change perspective advocates that changes in international law may be necessary to allow NLWs to be used as required for military and humanitarian reasons. This perspective posits that existing international legal rules affecting NLW development and use may be too restrictive and that blind obedience to such rules serves neither military necessity nor humanitarian principle.

In terms of NLW development, prohibitions on the development of 'non-lethal' chemical and biological weapons have already come under critical scrutiny by NLW proponents. In 1994, with respect to the CWC, Janet Morris argued that 'the writers of the treaty surely never intended to force us to kill people when we have an option to disperse people without killing them ... no one had any intention of prohibiting the development of environmentally friendly and humane non-lethals under the CWC'.[24] In 1995, two Air Force officers argued that 'current treaties must be renegotiated to take into account ... non-lethal technologies. Certain chemical and biological uses of non-lethal technology may be acceptable.'[25]

Influential experts in the US foreign policy establishment have also challenged the BWC and CWC strictures. The Council of Foreign Relations 1999 *Independent Task Force Report on Nonlethal Technologies* recommended that 'US security might be improved by a modification to a treaty such as the Chemical Weapons Convention or the Biological Weapons Convention'.[26] Russell Glen, an analyst at the RAND Corporation, has argued that 'chemicals can be our friends' in contemplating the development of NLWs for future US military operations in urban environments.[27] While expressing the belief that anything labelled 'biological warfare' will be banned, Alexander also encourages that 'non-lethal' biological weapons 'be evaluated on a case-by-case basis' rather than be subject to the existing prophylactic prohibition.[1]

Alexander has expressed similar selective change sentiments more broadly:

> Given the change in the nature of conflict in many areas, limited or indistinct objectives, and rapid advances in technology, treaties and conventions that have already been agreed to may need to be reexamined. That position will not be popular with organizations [e.g., International Committee of the Red Cross] that have established careers based on developing the agreements and enforcing them. A critical issue in future weapons system development should be a comparison of alternatives as they affect desired outcomes. If the laws turn out to be wrong, change them. We should not be tied to emotionally based, anachronistic circumstances.[1]

He did not elaborate on what rules of international law might need to be changed to accommodate more expansive use of NLWs but the position that international law may need to change to accommodate more robust use of NLWs is clear. The changes should be made selectively on a case-by-case basis rather than through a radical overhaul of applicable international legal rules.

The selective change perspective gives more moral and legal weight to the changes in the nature of warfare and military technologies than the compliance perspective. In other words, strictures on NLW development and use may prove anachronistic because they were developed with other concepts of warfare and military technologies in mind. This argument includes the perception that existing rules of international law developed to control traditional warfare fought with lethal weaponry. Alexander summarized this view by arguing that 'as conditions have changed, so must our legal imperatives'.[1] Russell Glenn echoed this view when arguing that the development of NLW capabilities may be outpacing international law.[27]

Restricting NLW development and use in the radically altered conditions of military operations today strikes some NLW proponents as morally bizarre because the only available options are lethal force or no military action at all, both of which are unrealistic. Changes in warfare and military technologies may mean that the relationship between international law and armed conflict is characterized less by the tension between military necessity and humanitarian principle created by lethal weaponry than by an alliance of humanitarian necessity and military principle created by the potential of NLWs.

The frustration experienced by international humanitarian lawyers and organizations in applying existing international legal rules to the new kinds of military operations support the selective change perspective. The Head of the Legal Division of the International Committee of the Red Cross identified difficulties with applying IHL to multilateral peacekeeping and humanitarian operations.[28] She referred to the problem of implementing rules designed for conflicts between belligerents in military operations other than war as 'insuperable' and she argued that international negotiations and co-operation are needed to find ways to adapt IHL to the changed circumstances faced by military forces.[28]

In sum, the selective change perspective posits that international law should, in certain situations, have different rules for NLWs than for lethal weapons. This perspective does not hold compliance with existing international law in high esteem because warfare and war-fighting technologies have changed and are changing. Other than calls for BWC and CWC modifications, it is not easy to identify what other rules of international law need to be modified to allow for more robust NLW use. Given the nascent development of NLWs, this ambiguity is understandable, as is the preference for a case-by-case approach to reform. More technological developments are needed before the selective change perspective can become more specific. However, the selective change perspective raises the need to define NLWs more objectively if they are to be subject to different international legal rules than lethal weapons and there are many reasons why such a definitional task may be difficult to achieve.[29]

Registering the Full Impact of 'Future War': The Radical Change Perspective

A third, more revolutionary, perspective lurks within the discourse on the impact of international law on NLW development and use. The radical change perspective emerges because neither the compliance nor the selective change perspectives register the full implications of what NLW proponents refer to as 'future war'. The compliance perspective emphasizes the importance of the continuity of existing international legal rules and rejects more lenient treatment for NLWs. The selective change perspective uses changes in military operations and technologies as a basis for advocating selective, case-by-case reforms in international law to allow more NLW development and use, but it does not embrace the more radical implications of 'future war'. Giving these implications more weight produces the radical change perspective.

In summarizing developments in the nature of warfare, Alexander observed that 'in many future military missions, ... use of deadly force will necessarily be minimized'.[1] Technology plays a central role in allowing militaries to minimize use of lethal force, whether the technology involved is 'smart' weapons, such as computer- or laser-guided weapons, or NLWs.

NLWs and 'smart' weapons differ, however, in important ways. 'Smart' weapons were primarily designed to allow military forces to destroy enemy military targets with lethal force and conform to the underlying moral and legal paradigm for regulating warfare. For example, 'smart' weapons reinforce the age-old moral and legal principles of discriminating between civilian and military targets. NLWs have a more radical aspect because their development and use challenges the traditional moral and legal perspectives on armed conflict. This radical edge of NLWs and how it affects international law is explored below.

'Future War' and the Christian Just War Theory

The radical change perspective contains a critical evaluation of the CJW theory. Arguments about the changed and changing nature of warfare and weaponry challenge assumptions embedded in the CJW theory that may no longer be either morally or legally credible. The CJW theory contains several key assumptions. First, it assumes that political leaders and states can agree on what limited situations qualify as 'just causes' for engaging in war; second, that organized, professional armies conduct hostilities; third, that warfare will be between states and, fourth, that separating combatants from non-combatants is possible in engaging in armed hostilities. The theory has a static view of military technology because changes in military weaponry do not affect the applicability of the moral principles.

NLW literature contains analysis that implicitly (if not explicitly) challenges these assumptions. First, NLWs may expand rather than limit the 'just causes' for using force, thereby reversing the trend in international law to restrict severely legitimate uses of force. The controversy surrounding humanitarian intervention might subside if military forces could deploy NLWs to intervene against atrocities and massive human rights violations perpetrated by rogue regimes. Anticipatory self-defence might be viewed more favourably if undertaken with NLWs rather than merely with lethal force. Attacks on terrorist groups harboured inside states might be less controversial if the attacks were conducted with NLWs. Using 'non-lethal' force to enforce economic sanctions may also become accepted.

Second, NLW literature is replete with assertions about non-traditional warfare that calls into question the attachment of CJW theory to the traditional hostilities conducted by states through professional armies. Non-traditional warfare, particularly in urban settings, makes separating combatants from non-combatants extremely difficult, if not impossible. As Herbert argued, 'conflict is shifting from the open battlefields to densely populated urban centers, making the combatant and noncombatant almost indistinguishable'.[30] This reality erodes the emphasis of CJW theory on distinguishing civilians from military forces. In addition, peacekeeping troops have often found civilian populations dangerous to their security and wellbeing because combatants use civilians as shields. Also, the civilians themselves sometimes present a threat to military personnel and equipment, but soldiers have been unable to respond to these novel threats with lethal force because of proscriptions against intentionally attacking civilians. NLWs can intentionally be used against civilian populations and targets but the CJW theory cannot support such a strategy.

Third, NLW literature contains a dynamic rather than static view of military technology. Arguments in favour of developing and deploying NLWs often rely on the new capabilities such weapons give military forces and suggest that such capabilities affect how we evaluate the ethics of weapons' use. The CJW theory, and the international law it informs, developed with a static technological perspective fixated on lethal force. Much of international law on the use of force and armed conflict in the twentieth century developed to restrain ever more lethal and destructive weaponry. NLWs change the technological paradigm radically in a way that ethical evaluation of 'future war' has not grasped.

The Radical Change Perspective and International Law on NLWs

In keeping with the radical change perspective's challenge to the CJW theory, its position on existing international law is equally dramatic. Table 2 summarizes the international legal changes suggested by the radical change perspective and presents them against the tenets of the CJW theory and existing rules of international law.

Space constraints prevent me from analyzing the international legal changes at which the radical change perspective hints. The most dramatic

TABLE 2
THE RADICAL CHANGE PERSPECTIVE IN PERSPECTIVE

Tenet of the CJW Theory	Manifestation in Current International Law	International Legal Rules Suggested by Need for NLWs
A leader may only engage in war for a just cause	A state is prohibited from using force in international relations, except in cases of individual self-defence, collective self-defence, or under authorization from the UN Security Council in response to a breach of international peace and security	NLWs may increase legitimacy for use of force in connection with humanitarian intervention, anticipatory self-defence, enforcing sanctions and dealing with threats from terrorist groups
Military force may not be used intentionally against civilians	Military forces are prohibited from attacking civilians and civilian targets	Military forces are allowed to use NLWs directly against civilian populations as part of military strategy

changes are widening the list of legitimate reasons for using force and allowing civilians to be legitimate targets of military 'non-lethal' force. Opening up these possibilities creates the need for applying the proportionality and 'unnecessary suffering or superfluous injury' principles to uses of 'non-lethal' force against civilians and, since the radical change perspective posits that direct use of NLWs against civilians is legitimate, collateral civilian effects from the use of NLWs against military targets cannot be considered illegal. Such a set of radical changes to international law would also affect arms control restrictions on conventional, chemical and biological NLWs, increasing the potential that such weapons may be developed and legitimately used.

Again, what I sketch above has not been advocated, to my knowledge, by any individual or organization. While it is speculative, real-world events contain glimpses of the radical change perspective. Soldiers' frustration with military operations other than war in part results because they have no weapons or tactics that work well in conflict zones where civilians are endemic. Much of the push for NLWs is to find weapons that can be used legitimately against civilians and civilian targets in mixed combatant/non-combatant environments. For example, Colonel George Fenton, Director of the US Joint Non-Lethal Weapons Directorate, indicated that he 'would like a magic dust that would put everyone in a building to sleep, combatants and non-combatants'.[31]

Further, if NLWs become more sophisticated and powerful, their potential may alter how experts look at the morality and legality of humanitarian intervention, anticipatory self-defence, enforcement of sanctions and attacks on terrorist groups. For example, NATO emphasized its use of a 'non-lethal' black-out bomb during its humanitarian intervention in Kosovo,[32] suggesting that NLWs can affect how people view the legitimacy of such military actions.

The radical change perspective contains multiple assumptions and problems that I do not have space to examine. The basic message I send in constructing this perspective is that the underlying military and humanitarian reasons for developing and using NLWs, combined with the successful development of NLW technologies, may have more radical implications for international law than has yet been discussed in NLW literature.

Conclusion

Their proponents almost universally argue that NLWs are needed if military forces are to address the changing nature of military operations. Pressure is building to accelerate NLW research, development and deployment. Conventional international legal analysis of NLWs does not, however, share this enthusiasm for NLWs because existing international legal rules ban the development of certain kinds of NLWs and restrict their military uses. Not surprisingly, NLWs partake of the age-old tension between military necessity and international law.

This chapter has presented three perspectives on what the impact of NLWs should be on international law. The compliance perspective resolves the age-old tension in favour of international law. The selective change perspective resolves the tension by advocating for more international legal accommodation of NLWs. The radical change perspective blows the tension apart because NLWs create the opportunity and need to reformulate moral and legal norms on the use of force and armed conflict.

It is too early to determine which perspective will prevail. The existence of three different perspectives suggests, at least, that the relationship between international law and NLWs will be more complex, controversial and dangerous than people may realize. In addition, the three perspectives indicate that the battle over which one will triumph in 'future war' has already begun.

Notes

1. Alexander J. *Future War: Non-Lethal Weapons in Modern Warfare.* New York: St. Martin's Press, 1999.
2. Coppernoll MA. The Non-Lethal Weapons Debate. *Naval War College Review* 1999; 52: 112–31.
3. Dando M. *A New Form of Warfare: The Rise of Non-Lethal Weapons.* London: Brassey's, 1996.
4. Fidler DP. The International Legal Implications of 'Non-Lethal' Weapons. *Michigan Journal of International Law* 1999; 21: 51–100.
5. Lewer N, Schofield S. *Non-Lethal Weapons: A Fatal Attraction.* London: Zed Books, 1997.
6. Hoffmann S. *Duties Beyond Borders.* Syracuse: Syracuse University Press, 1981.
7. Walzer M. *Just and Unjust Wars.* New York: Basic Books, 1977.
8. United Nations Charter, 1945. In: Brownlie I, ed. *Basic Documents in International Law.* Oxford: Oxford University Press, 1995, 4th Edn: 1–35.
9. Geneva Protocol, 1925. In: Roberts A, Guelff R, eds. *Documents on the Laws of War.* Oxford: Oxford University Press, 1989, 2nd Edn: 139–40.
10. Biological Weapons Convention, 1972. In: *International Legal Materials* 1972; 11: 309–15.
11. Protocol IV on Blinding Laser Weapons, 1995. In: *International Review of the Red Cross* 1995; 312: 272–99.
12. Land Mine Convention, 1997. In: *International Review of the Red Cross* 1997; 320: 563–78.
13. Chemical Weapons Convention, 1993. In: *International Legal Materials* 1993; 32: 800–73.
14. Additional Protocol I, 1977. In: Roberts A, Guelff R, eds. *Documents on the Laws of War.* Oxford: Oxford University Press, 1989, 2nd Edn: 389–446.
15. Geneva Convention I, 1949. In: Roberts A, Guelff R, eds. *Documents on the Laws of War.* Oxford: Oxford University Press, 1989, 2nd Edn: 171–92.
16. Lewer N. Nonlethal Weapons. *Forum for Applied Research and Public Policy,* 1999: 39–45.

17. St. Petersburg Declaration, 1868. In: Roberts A, Guelff R, eds. *Documents on the Laws of War.* Oxford: Oxford University Press, 1989, 2nd Edn: 30–31.
18. Hague Convention IV, 1907. In: Roberts A, Guelff R, eds. *Documents on the Laws of War.* Oxford: Oxford University Press, 1989, 2nd Edn: 44–57.
19. Conventional Weapons Convention, 1981. In: Roberts A, Guelff R, eds. *Documents on the Laws of War.* Oxford: Oxford University Press, 1989, 2nd Edn: 471–79.
20. Coupland RM. 'Non-Lethal' Weapons: Precipitating a New Arms Race. *BMJ* 1997; 315: 72.
21. Environmental Modification Techniques Convention, 1977. In: Roberts A, Guelff R, eds. *Documents on the Laws of War.* Oxford: Oxford University Press, 1989, 2nd Edn: 379–86.
22. NATO. NATO Policy on Non-Lethal Weapons. Press Statement, 13 Oct. 1999.
23. US Department of Defense. *Directive No. 3000.3: Policy for Non-Lethal Weapons* 9 July 1996.
24. Lewer N. Non-Lethal Weapons. *Med War* 1995; 11: 78–95.
25. Klaaren, JW, Mitchell RA. Nonlethal Technology and Airpower: A Winning Combination for Strategic Paralysis. *Airpower Journal* (Special Issue) 1995: 42–51.
26. Council on Foreign Relations. *Independent Task Force Report – Nonlethal Technologies: Progress and Prospects*, 1999: <http://www.foreignrelations.org/public/pubs/Non-ViolentTaskForce.html>
27. Glenn R. Separating the Wheat and the Chaff: Non-Lethal Capabilities in Future Urban Operations. Paper presented at Jane's 4th Annual Non-Lethal Weapons 2000 Conference, 5 Dec. 2000.
28. Doswald-Beck L. Implementation of International Humanitarian Law in Future Wars. *Naval War College Review* 1999; 54: 24–52.
29. Kenny J. Potential Health Effects of Non-Lethal Weapons. Paper presented at the First Annual Non-Lethal Technology and Academic Research Symposium, Quantico, Virginia, 3–5 May 1999.
30. Herbert DB. When Lethal Force Won't Do. *Proceedings* Feb. 1998: 47–9.
31. War without Tears? Should 'Non-Lethal' Chemical and Biological Weapons Be Allowed? *New Sci* 16 Dec. 2000, at <http://www.newscientist.com/news/news.jsp?id=ns22693>
32. Rosenfeld SS. Turning Off the Lights in Belgrade. *Washington Post*, 7 May 1999: A39.

The Revolution in Military Affairs Debate and Non-Lethal Weapons

GERRARD QUILLE

Strategy and Politics

Strategy

When considering strategy it is useful to be reminded of what we are talking about and the parameters of discussion. Strategy is of course a contested concept. However, definitions abound and the British/American strategist, Colin S. Gray, has recently written that:

> The complexity of strategy and war – conflict on land, at sea, and in the air, and in space and cyberspace – is modest compared with the complexity of the dimensions, factors, or elements that interactively comprise their nature.[1]

He describes 'seventeen dimensions' that are clustered into three categories. The first category, 'People and Politics', comprises people, society, culture, politics and ethics. The second category, 'Preparation for War', includes economics and logistics, organization (including defence and force planning), military administration (including recruitment, training and most aspects of armament), information and intelligence, strategic theory and doctrine and technology. The final category, 'War Proper', is composed of: military operations, command (political and military), geography, friction (including chance and uncertainty), the adversary and time.[1]

It is this holistic approach to understanding the complexity of modern conflict and the formulation of strategy that is necessary, according to Gray. He warns that 'Strategy is seriously incomplete if considered in the absence of any one of them'. Gray argues that the preference for the technological persuasions of the revolution in military affairs (RMA) should be put in the context of such dimensions because, by its nature, the RMA leads 'to persuasion by unsound theories of miracle cures for strategic ills'.[2]

This approach to understanding strategy is similar to Michael Howard's four dimensions – social, logistical, operational and technological.[3] Both authors draw heavily on Clausewitz who described a typology of strategy in the 'elements of strategy' which includes moral, physical, mathematical, geographical and statistical aspects. This is important because RMA discussions, as will be outlined below, often emphasize one dimension, the technological, to the detriment of the complexity of strategy and modern conflict itself. Thus one question for consideration might be whether RMA discussions reflect the complexity of strategy and conflict or whether they over-rely on the importance of technology?

Politics as the Framework for Strategy

Although Clausewitz (in his reference to *politi*) and Gray (in his reference to strategy as the bridge between military power and political purposes) both refer to the domain of the political, all too often the discussion of strategy (particularly in the context of RMA) neglects the framework of politics, in this case international politics.[1] Even though Gray can tell us that the United States dominated the last century and that it will continue to do so into the next century, he does not tell us how this relates to the US' role in the nature of conflict. Paul Rogers describes the international system in terms of 'a violent peace' where the US and strong western allies can maintain 'a kind of peace and order' that assumes present patterns and attempts to keep them contained, *liddism*, rather than one that tries to understand and work towards their resolution.[4] This is central to the political context of the RMA and to the question of whether we intend to maintain a violent peace, or to ask whether other strategies or political futures are more desirable than the RMA's 'military futures study'.[5]

A discussion of strategy alone does not help the policy-maker or the doctrine writer, both of whom also need to understand the political and social environment they are working in and the nature of war and present conflict. As Air Marshal Sir Timothy Garden pointed out in a recent review of Gray's *Modern Strategy*, 'whilst Gray points out the eternal nature of strategy he does not tell us anything about modern conflict'.[6] If debates on the RMA are to have any meaning in the framework of international politics, they must be situated in an understanding of conflict environments as the means for pursuing the international communities' legal and moral obligation to seek the peaceful resolution of such conflicts.

This provides the second introductory point for this chapter. When considering the concept of RMA one must also be aware of the political, in this case the international political, and security environment when moving from concept to strategy and then to operational and tactical deployment. This raises the question of whether the RMA discussions refer to the trends of contemporary conflict in the international system or to imagined 'military' future conflicts. That is, whether the RMA debate is strategic or political.

If the RMA is seen only as an exercise in military future studies that exclude the complexity of strategy and conflict or present and future projections of the international political and security environment, its limitations should be recognized when considering its application to policy and doctrine.

The Revolution in Military Affairs

There are now many historical references to RMAs. The more one looks at each century the easier it is to find examples when technological developments and organizational and operational developments combine to create a new system of warfare or revolution in military affairs. One historical investigation noted that the first application of the term 'military revolution' dates at least from Michael Roberts' 1955 lecture on the Swedish adoption of massed rifle volleys in the sixteenth and seventeenth centuries; this raised Sweden's international status to a degree disproportionate to its demography and resources.[7]

History is littered with further instances of major shifts in military organization and strategic rationale that resulted in substantial changes in regional or international status and which could therefore be regarded as 'military revolutions'. There have also been a number of developments in the twentieth century which have significantly contributed to the lexicon of RMA. For example:

- The German *Blitzkrieg* during the Second World War (see below).

- The Soviet Union's concept of a Revolution in Military-Technical Affairs (RMTA), which, in the 1960s, foresaw the potential future application of the nuclear revolution. In the 1980s, the potential of the Strategic Defence Initiative (SDI) to exploit Soviet fears of Russian technological and economic inadequacy *vis-à-vis* the US, for example, the development of new remote space-based satellite technologies to intercept and destroy nuclear ballistic attacks against the US was recognized.

- The US Military Technical Revolution (MTR) perceived that overwhelming Soviet conventional forces were to be offset with 'technological' advances that enabled NATO to strike deep and accurately at Soviet follow-on forces.

- Spin-offs from the nuclear and 'space-race' programmes, such as precision guidance, ballistic missiles, electronic processing systems and the like, which have provided the basis of today's archetypal RMA technologies. Spin-off should not be read as accidental; this category includes Research and Development (RD) on military technology, stemming from radar and nuclear weapons research, which significantly forced the pace of innovation.

Defining RMA in the Present Debate

RMAs have been defined as occurring:

> when new technologies (internal combustion engines) are incorporated into militarily significant number of systems (main battle tanks) which are then combined with innovative operational concepts (Blitzkrieg tactic) and new organisational adaptation (Panzer divisions) to produce quantum improvements in military effectiveness. The twentieth century is marked by three military revolutions: mechanised warfare in the 1930s and 1940s; nuclear weapons and ballistic missiles in the 1950s and 1960s; and cybernetics and automated troop control (information technology) beginning in the 1970s and continuing into the twenty-first century.[8]

The absence of any one such characteristic could make the difference between a technological or strategic development having a revolutionary impact. For example, although the French tank divisions before the Second World War were very good, they lacked the organizational capability that was the key to the success of the German *Blitzkrieg*.

The acknowledgement that the present RMA stems from the 1970s and will run into the twenty-first century raises the question of whether this process is *evolutionary* rather than *revolutionary*. Whether we are in fact witnessing revolutionary change socially and politically, let alone militarily, remains a moot point. This suggests caution in applying premature conclusions to the under-studied relationships between the ongoing political, social and military changes and the nature of conflict.

What is Information Warfare?

RMA technologies can be divided into two groups: first, the 'hard technologies' including 'smart weapons' and sophisticated weapons platforms, second, the information technologies that enable the functioning and accuracy of 'smart weapons' and the integration of Command, Control, Communications, Computers and Intelligence (C_4I_2) in sophisticated weapons platforms.

Information has undoubtedly had a significant impact on the way in which warfare is conducted and how its outcome is shaped. Examples include information interception, such as that achieved at Bletchley Park during the Second World War; information-manipulation, such as *Bismarck*'s use of William I's telegram in the 'Dispatch of Ems' to unify support for war with France; and information for action, such as Eisenhower's decision to commence with D-Day based on his weather forecasters' predictions. Innovations in technology and process have allowed an increase in the use and accuracy of information as a weapon of war.

> One important form of information warfare is decision-making warfare, in which a defender or attacker uses information acquisition or processing technology to complete their decision-making cycle quicker than an opponent can to maintain the initiative in the battle.[9]

Disrupting or defending the decision-making process to achieve 'Information Dominance' is the goal. It is imperative to process decisions quicker than an adversary. The 'real-time' aspect of this information acquisition and processing technology 'is creating a revolution in the way military operations are conducted'. However, agreement over the exact form of this revolution remains contentious.

In its purest sense, IW involves a radical shift away from the traditional confrontation of mass armies to conflict 'behind the lines', aimed principally at the adversary's infrastructure and leadership. However, this is far from the present reality. For example, the Gulf War was a hybrid version of a conventional military operation in which information warfare (IW) was used as a partial, but not the sole, means of achieving military objectives.

RMA and the US Debate

In the US the RMA appears to have become the most significant driving-force for defence policy, such that we see 'the executive and legislative branches eagerly, even if at times ignorantly, urging on a military already driven by its own inexorable organisational impulses'.[7]

The presentation of future wars conducted from a distance with stand-off weapons and satellite remote sensors, often associated with the RMA, must be attractive to US political elite. 'Full of promise, it [RMA] seems to offer Americans an answer to many enduring strategic dilemmas, whether intolerance of casualties, impatience, or the shrinking military manpower base.'[10] This is echoed by Gray's assessment of US cultural bias whereby:

> Strategy and war are holistic enterprises. US strategic culture is wont to function taking one thing at a time on its own merits. Monochronic defense performance leads to a focus on only one or two dimensions of what is almost always a more complex challenge.[2]

It is a basic premise of RMA purists that information and its technologies are the key to power status and success in future warfare. Although information itself cannot win a battle, it can influence whether or not and where a confrontation takes place. This has fuelled concern that the RMA may be elevated to the status of a doctrine.

The Legacy of the 1991 Gulf War

The Gulf War is essentially the departure point for advocates and detractors of RMA alike. The speed, accuracy and lethality of the weaponry and the management of the multi-dimensional battlefield excited, impressed and astounded those watching real-time media despatches. We were told that the Allies rendered Iraq impotent by incapacitating its communications. Such was the apparent success of the new technologies and the disabling of Iraqi command and control, that Alan D. Campen (former Director of Command and Control Policy in the US Defence Department), wrote that in the Gulf War:

> knowledge came to rival weapons and tactics in importance, giving credence to the notion that an enemy might be brought to its knees principally through destruction and disruption of the means for command and control.[11]

Information domination was complemented by domination of the air, which resulted in a quick and decisive land-campaign. These features were viewed as indicative of how future wars would be conducted, that is, in a three-dimensional conflict zone where traditional lines of conflict are replaced by multi-dimensional real-time conflicts targeting an enemy's C_4I_2 in the rear as well as engaging his troops with both long-range, precision-guided missiles (incapacitators) and rapid, close counter-attacks. In the Gulf this was achieved with cruise missiles, smart bombs and stealth technologies.

Some analysts claim that such technologies will allow future wars to be fought from greater distances without the need for massing at theatre entry points, thus removing the tell-tale signs of an impending strike and thereby reducing the likelihood of casualties and collateral damage. However, not everyone has been convinced that the Gulf War represents a model for future conflict or that it justifies elevating the RMA above other factors.

It was reported that Iraqi ineptitude did not require the Coalition to really test new technologies such as the Joint Surveillance Target Attack Radar System (JSTARS). However, the authors did investigate five technologies used during the military operation (including stealth/low observability laser-guided bombs, air-refuelling, the high-speed anti-radiation missile and the Secure-Telephone Unit [STU] III). They concluded that 'For the most part, these technologies were not really new and were available in less sophisticated forms during the Vietnam War'.[12] Their conclusions raise serious questions about the revolutionary nature of the Gulf War and subsequent careless assertions about revolutionary doctrines based on the promise of technology. The RMA's influence was of a hybrid rather than purist nature. After all, 'some of the aspects of the war that seemed most dramatic at the time appear less so than they did in the immediate afterglow of one of the most one-sided campaigns in military history'.[13]

Asymmetrical Warfare

Arquilla warned of the 'diffusion effects' of imitation.[7] Whilst the present RMA-technologies represent a great challenge to would-be imitators, it is not an insurmountable one. Yet even if states outside the transatlantic community do develop limited RMA-type technologies, the strategic balance will still reflect the dominant role of the US as the sole superpower. It is inconceivable that any state could seriously challenge US superiority in a Major Regional Contingency (MRC) during the foreseeable future.

The costs associated with adopting large, technologically advanced military platforms and integrating them with C_4I_2 systems in order to counter (or keep up with) the US and its allies may prompt some states to conclude, as did an Indian General, that the main lesson of the Gulf War was never to fight the US without nuclear weapons.[14] Whilst a nuclear programme is not only challenging financially and technically, chemical and in particular biological weapons are more affordable and less technically challenging alternative weapons of mass destruction (WMD). Future adversaries are more likely to pursue the current trend away from the concept of a 'decisive battle'.

RMA and Weapons of Mass Destruction

The link between the RMA and weapons of mass destruction (WMD) is detectable in discussions within US policy-making circles. As the US National Defense Panel concluded:

> WMD will require us to increase dramatically the means to project lethal power from extended ranges. We must provide a conventional, non-nuclear deterrent capability against the use of weapons of mass destruction.[15]

Yet RMA as a counter-force to WMD raises as many problems as it solves. For example, during the recent stand-off between the US and Iraq it was acknowledged that precision-guided smart bombs would have been unable to target and successfully destroy all of Iraq's known biological capabilities. More research following the Gulf War is raising questions about the impact of Iraq's known chemical and biological weapons (CBW) programme in deterring the US from pursuing Saddam Hussain to Baghdad and intervening in his actions against the Kurds immediately following the Gulf War.[4]

Lawrence Freedman, who has examined RMA in the context of a 'Western Way of War', acknowledges that there has been a shift away from the 'decisive battle' to a 'process of destruction'[16] and that:

The effort to reverse this tendency, so that war can again become a fight, is the core theme of much of the strategic theory of the past quarter-century, even in the nuclear sphere. The RMA represents the culmination of these efforts.[17]

Nuclear weapons have had a major influence on shaping this situation:

From Central Europe to Kashmir, and from the Middle East to Korea, nuclear weapons are making it impossible for large sovereign units, or states, to fight each other in earnest without running the risk of mutual suicide.[18]

In comparison to Iraq's biological warfare programme with its relative ease of development, production and storage undetected by the international community, the RMA rationale for conventional war appears very risky.

The pre-occupation of RMAs with MRCs also neglects current trends towards intra-state rather than inter-state conflicts. As the *Stockholm International Peace Research Institute (SIPRI) Annual Yearbook* concluded: 'All but one of the conflicts recorded for 1996 were internal, that is, the incompatibility concerned control over the government or the territory of one state.'[19] What then are the most likely conflict scenarios and what provisions does RMA afford them?

Conflict Short of War

Compared to MRCs, some analysts conclude that 'With a few exceptions the impact of the RMA on conflict short of war is far less clear'.[1] Whatever the technological merits of large and expensive military hardware, the fact remains that in many present operations involving US and UK troops, such as United Nations Peace Support Operations (PSOs), the human element of troops in the field is essential to their success. UN peacekeepers on the ground represent a visible, as well as political, commitment to resolve a conflict. This would be very difficult to replace through long distance force projection. Once again, a balance is required to integrate technological advances into doctrines designed for such conflict scenarios as PSOs. Here, the RMA is no substitute for doctrine.

Many analysts point out that weapons designed for High Intensity Conflicts (HIC) can be equally effective in Low Intensity Conflicts (LIC), for example, the peace-support role of British Challenger tanks in Bosnia. However, certain capabilities bracketed with the RMA, such as stand-off precision weapons, have had their value questioned in conflicts short of war.[10] Another concern is that the RMA debate may overshadow the needs of soldiers in zones of conflict that do not rely just on technology but on 'contact skills' that are essential for their success in peacekeeping operations.

During the cold war a peacekeeper was traditionally conceived of as having an inter-positionary role. Today the role of a peacekeeper has been broadened to cover Peace Support Operations (PSOs) including inter-positionary roles as well as peace-enforcement. The broadening of the peacekeeper's role has highlighted the need for weapons and weapons systems (both for self-defence and for peace-enforcement, in other words fighting) and the need for the development of contact skills for peacekeepers on the ground (to support their *de facto* mediation and negotiation activities). In the context of the RMA and NLWs debates, the focus is on weapons and weapons systems for peacekeepers, rather than on the less developed focus on the day-to-day needs of soldiers on the ground, that is, contact skills.

It is in the niche market for a spectrum of war-fighting capabilities and platforms applied to PSOs in urban environments that the debate on the RMA and NLWs are attracting the attention of policy-makers and doctrine writers. This represents the bias of weapons-based debates on the RMA and NLW, and their military utility, which overlooks other important needs of soldiers today.

The Role of Technology in Conflict

The recent Carnegie Report, *Preventing Deadly Conflict*, acknowledged the impact of technology in today's society and observed that:

> Historically, technological advances have resulted in social and economic transformations on a vast scale. This is especially likely when fundamental new technologies are unfolding across the entire frontier of scientific and engineering research and are rapidly disseminated throughout the world. The impact over the longer term has been positive. Along the way, there have been massive dislocations. In this context, it is worth recalling the severe disruptions of the industrial revolution; they had much to do with the emergence of communism and fascism, especially the Nazi catastrophe.[20]

The main problem today, as the Carnegie Commission emphasizes, is not necessarily identifying potential areas of conflict but actually acting on that information. Co-ordinating activities quickly and decisively could minimize Bosnia, Rwanda and Burundi-type conflicts degenerating to the levels they did. Such conflicts also highlight the deficit between information and action at a political level which contrasts with RMA debates on military integration for information systems and military systems for rapid response.

Technological advances have played a part in peacekeeping situations, such as the success of tele-medicine in Bosnia, where peacekeepers and civilians benefited from the 'primary diagnosis of radiology exams performed by radiologists stationed in Hungary and Germany'.[21]

We must avoid adopting the slogan of the RMA in place of a doctrine because of a misplaced desire to enhance the credibility of expensive military platforms that may not significantly contribute to present and probable future conflicts. As Freedman warns, 'It will ... hardly be a revolution in military affairs if it leads those who embrace it to avoid most contemporary conflicts, and only take on those that promise certain and relatively painless victories'.[22]

PSOs in densely populated urban environments offer a greater and more urgent challenge to military and political decision-makers than the prospect of large armies face-to-face in large open spaces, such as the desert in the Gulf. Urban warfare environments represent a pressing area for discussion by military planners because, in support of the SIPRI figures above, the world's population is becoming more urbanized. It has been estimated that approximately 44 per cent of the total world population will reside in urban areas by the year 2005. Examples in an operational context are the strategic importance of Grozny in Chechnya, the towns and cities of Kosovo for NATO after the air war and Freetown in Sierra Leone for the British recently. Nevertheless, the preference of RMA proponents to focus on MRCs reflects the observation that urban environments are amongst the least preferred operational settings. The difficulties of an urban environment range from the terrain to the problem of distinguishing civilians and combatants. In such environments technological advances in sensors and reconnaissance will be welcome but the application of NLWs for urban warfare is in a preliminary stage and the implications of it are far from clear in operational and legal senses.

The RMA and Implications for the UK

The RMA debate can be considered as a derivative of the US political and defence identity. The combination of an inherent abhorrence of casualties, along with a tradition of seeking major technological breakthroughs (specifically military ones), has propelled the current RMA beyond concept and debate and toward its present doctrinal heights.

However, NATO allies should have some concern about the implications of the RMA debate in the US. If the US sees itself as the sole supplier of heavy-lift capabilities (a reality at present) and certain command and control systems including stand-off and Precision-Guided Munitions (PGMs), how does it perceive its allies' role *vis-à-vis* such technologies? Do US policy-makers now envisage their forces remaining dislocated from the potential bloody battlefield? Instead, will the US facilitate allied forces in the field, using its command and control systems backed up by long-range surgical strikes, whilst the less technologically endowed allies role up their sleeves and engage the adversary in the battle-zone? In other words, the US will play an anaesthetic role with its allies doing the close-up and/or clean-up operations. Allies are unlikely to regard such a division of labour as politically acceptable.

Will a United States-RMA leave its Allies Behind?

The UK's 1998 Strategic Defence Review (SDR) acknowledges the possibility of the US leaving its less technologically endowed Allies behind by asking: 'How do we and our Allies retain interoperability with US forces given the radical changes they envisage?'[23] On the other hand, the US will not want to deploy any technological systems or platforms that impair co-operating with its allies. The White Paper asks 'will technological changes also require radical changes in the way our forces are organised and fight?'

RMA Phase 2: A Role for Non-Lethal Weapons?

The prospect of NLWs as incapacitators of humans and machinery is attractive when considering the operational problems of urban warfare environments.[5] Ullman and Wade[24] outline how such a system might develop and be integrated with NLWs through a strategy entitled 'Rapid Dominance – A Force for All Seasons':

> Since 'rapidity' means exactly that, we sought to provide a range of nearly instantaneous responses for the president, capable of delivering lethal or non-lethal weaponry as well as other significant actions to affect the will and perception of an adversary.[24]

They envisage that such a concept could be operational in 15 years including research and development time-scales. The applications of the strategy of rapid dominance applying 'shock and awe' would be:

> designed to be employed in a series of unrelenting 'waves' of powerful strikes across many targets combining sea air, land and space forces to affect and influence an adversary's will and perception independently of whether or not US and allied forces are forward deployed. Rapid Dominance includes, however, the capacity for the physical capture and occupation of territory should that be required.[24]

Military interest in incapacitants is not new. However, weapons such as anti-personnel mines (APMs) and cluster munitions have attracted adverse publicity that may have contributed to the attractiveness of the so-called non-lethal technologies. To this is added the trend toward urban conflict environments that military thinkers perceive to be important and the high likelihood of urban warfare becoming more common. The potential for non-lethal or smart technologies that can differentiate between civilian and adversary is another driver.

The direct relationship between larger-scale RMA technologies (communications and weapons platforms and integrated through systems) and NLWs has been identified as the beginning of what could be a second radical phase of the RMA.[5] This phase would include far more advanced technological applications to future warfare. Its developments include the following stages:

- *Cyber-warfare and 'strategic information warfare'* – Increased use of networks and information connections.

- *Robotics* – Essentially about the integration of robotic networks with miniaturization such as micro-electromechanical systems or MEMS (Darpa). Use of tiny mechanical devices coupled to electrical sensors and actuators, when combined with nano-technology, could lead to bio-mimicry developments such as a robotic tick to gather information.

- *Cyborgs* – Simple cyborgs such as cameras and sensors on rats controlled by some kind of implants. This could be the beginning of a more complex form whereby developments in molecular biology, nano-technology and information technology will be combined and could lead to 'things like biological warfare weapons that are selective in targets and are triggered only by specific signals or circumstances'.

- *'Pyschotechnology'* – Technology to alter beliefs, perceptions and feelings of enemies.[5]

This integration of military operational needs in urban environments with NLW deployment is being given serious consideration by the RAND Arroyo team. Questions still remain about the conflict environment envisaged, whether war-fighting or enforcement rather than peacekeeping, and about the deployed weapons where surveillance is a traditional approach but chemical or biological incapacitants are new. On the latter, the implications of deployment against existing legal provisions such as those under the Biological and Toxin Weapons Convention of 1972 must be addressed. Furthermore:

> Beyond technological obstacles, the potential for effective battlefield robots raises a whole series of strategic, operational, and ethical issues, particularly when robots change from being lifters to killers. The idea of a killing system without direct human control is frightening. Because of this, developing the 'rules of engagement' for robotic warfare is likely to be extraordinarily contentious.[5]

Concern also arises about the reliability of such systems, given that they emphasize the need for speed and lethality, not just in distinguishing between civilian and combatant but also with regard to the margin of error or transparency in decision-making. This has implications for what that might mean for an 'info' attack to trigger such weapons without human input or oversight.

The combination of political, strategic, legal, social and proliferation concerns along with questions about the real needs of peacekeepers on the ground, should be enough to urge caution in any further movement from concept to operational provisions without careful and thorough debate.

Conclusion

If this debate is to be fruitful, those interested in NLWs must be aware of the direction of the RMA debate and some of its most contentious aspects. There is an increasing need to include a politico-strategic understanding of the implications of applying NLWs in what is already a problematic approach covered by the RMA.

If we are to clarify the emerging trends in conflict to help enhance a stable and sustainable security environment this century, the politico-strategic aspects of the debate should be re-opened in order to understand the objectives of the international community. This will avoid becoming locked into a military-technology-driven exercise in future studies.

The distinction between the two debates is central to the political context of the RMA. Do we intend to maintain a violent peace or are other strategies or political futures more desirable than the RMA's 'military futures study'?[5] The distinction is important and will determine the pattern of relations between great power states in both north and south and groups within areas of conflict (conflict complexes). The RMA promises to support an approach to conflict that emphasizes keeping the violent peace, that is, *liddism*. On the other hand, a broader political understanding of the nexus of conflict, development and societal security offers an understanding that demands shorter- and longer-term thinking about the path towards individual security, regional stability and a stronger international community. Neither approach will see the end of conflict nor of military intervention in the short-term but the latter offers the deeper understanding of conflict that is necessary to move towards its transformation, rather than its costly continuation, in both human and economic terms.

In summary, when considering the role of non-lethal weapons for the military in present and future conflicts, the concerns surrounding the 'heavier and organizing' end of the military-technology debate, the Revolution in Military Affairs, must be taken into account in their development.

Notes

1. Gray CS. *Modern Strategy* 2000: 6–7.
2. Gray CS. RMAs and the Dimensions of Strategy. *Joint Force Quarterly* Winter 1997–98; Autumn: 54.
3. Howard M. The Forgotten Dimensions of Strategy. *Foreign Affairs* 1979; 57: 976–86.
4. Rogers P. *Losing Control.* London: Pluto Press, 2000.
5. Metz S. The Next Twist of the RMA. *Parameters* 2000; **XXX** (3): 40–53.
6. Garden T. Book Review: Modern Strategy. *Times Higher Education Supplement*: 18 Feb. 2000.
7. Arquilla, J. The 'Velvet' Revolution in Military Affairs. *World Policy Journal* Winter 1997/98: 5.
8. International Institute for Strategic Studies. *Strategic Survey.* London: Oxford University Press/IISS, 1995/96: 30–31.
9. Perusich, K. Information Warfare: Radar in World War II as a Historical Example. *International Symposium on Technology and Society 1997*, IEE, 20–21 June 1997, University of Strathclyde, Glasgow: 92.
10. Metz S, Kievit J. *The Revolution in Military Affairs and Conflict Short of War.* Carlisle: US Army War College, 1994.
11. Toffler A, Toffler H. *War and Anti-War: Survival at the Dawn of the Twenty-First Century.* London: Warner Books, 1994: 87.
12. Kearney T, Cohen E. *Gulf War Air Power Summary Report.* Washington, DC: US Government Printing Office, 1993.
13. Freedman L. *The Revolution in Military Affairs.* London: International Institute for Strategic Studies/Oxford University Press, 1998: 45.
14. Manning RA. The Nuclear Age: The Next Chapter. *Foreign Policy* Winter 1997–98; 109: 71. Quoted in Note 12: 45.
15. Report of the National Defense Panel. *Transforming Defense: National Security in the Twenty-First Century.* Dec. 1997: 42.
16. Liddell Hart B. *The Revolution in Warfare.* London: Faber & Faber, 1946. Quoted in Note 12: 45.
17. Note 12: 15.
18. Note 12: 194.
19. Sollenburg M, Wallensteen P. Global patterns of major armed conflicts, 1989–1996. In: *SIPRI Yearbook 1997: Armaments, Disarmament and International Security.* Oxford: Oxford University Press, 1998: 17.
20. Carnegie Commission on Preventing Deadly Conflict. *Preventing Deadly Conflict. Final Report.* New York: Carnegie Corporation of New York, 1997: viii–ix.
21. Nairm G. Nominated for IT 'Oscars'. *Financial Times*, 1 April 1998: 12.
22. Note 12: 28.
23. *The Strategic Defence Review.* Cm3999. London: HMSO, 1998.
24. Ullman HK, Wade Jnr. JP. *Rapid Dominance – A Force for All Seasons.* London: Royal United Services Institute, 1998: vii.

Towards an Understanding of Non-Lethality

BRIAN RAPPERT

'Non-lethal' weapons have emerged as possible key solutions to contemporary challenges facing military and police forces. The rise of military operations other than war, the implications of the revolution in military affairs and the continuing threats to domestic order are all said by some to require novel technological responses.[1] At the heart of the promise of non-lethal weapons (NLWs) is the possibility of minimizing pain and death by providing flexible response tools for policing and military purposes. While the concept of non-lethal force is not new, particularly to law enforcement agencies, it has undergone something of a revival in recent years. Somewhat established kinetic, chemical and electric technologies are being accorded a heightened role, while scientific and technical expertise are being mobilized to design novel systems. New disabling calmative inducing agents offer possibilities for crowd control that are currently unrealizable, while magnetic and acoustic guns may replace old-fashion plastic bullets. Table 1 provides a partial listing of the wide range of current and future non-lethal technologies. Still, news of the present or future deployment of such weapons has not been met with universal welcome. Various civil liberties and human rights organizations, among others, have expressed doubts as to whether such developments simply give a high-tech wrapping (and perhaps legitimatization) to new ways of inflicting pain and suffering.

To believe the claims of proponents or critics though, the nature of policing and warfare may alter dramatically in the near future, in part because of the capabilities of non-lethal weapons. In this situation the pressing question arises of how to respond to such developments. The dilemma arises for those involved in peace research of how to respond to weapons ostensibly designed to reduce suffering. In what ways, for instance, might the effects of NLWs be deemed unacceptable? How could their use lead to dubious practices? What measures must be taken to ensure their positive aspects are realized? Any response to such questions must contend with a variety of complicating factors: the diversity of police, incarceration and military origins and contexts of use of NLWs, the plethora of relevant international and domestic laws that might be brought to bear on their deployment and the competing criteria advanced for their evaluation.

TABLE 1
TYPES OF NON-LETHAL WEAPONS

Category	Type	Description
Kinetic	Bean Bags	Non-penetrating sacks filled with lead shot
	Baton Round	PVC cylinder aimed at targets
	Rubber Bullets	Cylinders aimed at ground to ricochet off
Chemical	Gases	Irritant gas with chemical agent (e.g., CS, CN, CR, OC)
	Sprays	Hand-held spray with active agent
	Superlubricants	Prevents traction
	Superadhensives	Sticks materials
	Supercorrosives	Degrades materials
Electro-shock	Stun Baton	Baton which delivers electrical shock
	Stun Belts	Devices worn which deliver remotely triggered shock
	Taser Gun	Shoots two charged wire-trailing darts
	Water Stream	Electrified water stream
Acoustic	Infrasound Beam	Low frequency sound which causes disorientation, nausea
	Squawk Box	Ultrasonic device producing nausea in individuals
Optical Munitions	Visible Light Radiators	Causes flash blinding
	Strobe Lights	High intensity strobes which cause vertigo, disorientation
Lasers	Laser	Variable intensity; destroys optics, blinds, or 'dazzles'
Biological	Biodegrading Agents	Degrades materials
Other	Water Cannon	High pressure water stream
	Entanglers	Ensnaring objects; mine, gun, or other technology based
	Sticky Foam	superadhesive for immobilising objects
	Calmatives	Sleep-inducing chemicals
	Carbon Filaments	Disrupts equipment
	Aqueous Foams	Creates barriers, can be used with chemical irritants

While any number of approaches and substantive topic areas about NLWs might be addressed (e.g., the growing international trade in this area, the interacting and converging activities of the police and military), this chapter focuses on how we should conceive of and evaluate their effects. Initially we might expect that in the case of weapons their effects would be well established and such information could act as an agreed baseline for judgements about acceptability. Yet, this is far from the case. Fundamental differences exist in the basic characterization of NLWs. The next section maps the complications surrounding the notions of lethality and non-lethality as this relates to the design, deployments, contexts of use, objectives and moral considerations involved in these weapons. Several frameworks for regulating the use of NLWs are subsequently examined and critiqued in the section that follows. The case is made that forming such a scheme necessitates addressing the basic issue of how it is possible to ascribe particular characteristics to technology. In an attempt to build on as well as go beyond some of the limitations of these frameworks, a generalized approach is proposed for elucidating some of the complex effects of NLWs. The advantages of the framework offered is that it presents a forward-looking perspective that is sensitive to the range of claims made about the effects of weapons.

Lethality and Non-Lethality

An underlying conception of effects is central to the definition of non-lethal weapons. The term typically refers to weapons designed and deployed in a manner to minimize causalities and general long-term physical damage.[2] It is because of this 'minimization' that NLWs may help bring about a new manner of conflict resolution.

At first glance, it should be fairly straightforward to attribute consequences to particular technologies that would allow for their classification as lethal or non-lethal.[3] Weapons are designed to transfer a certain amount of energy to bodies thereby causing the shattering of bones, the tearing of flesh, or some other such impact. Non-lethal weapons are supposed to keep such damage to a minimum, yet the situation is not so clear. Plotting the probable effects of weapons is problematic. At one level this derives from an uncertainty over the range of devices that should be included for consideration as non-lethal. Within the literature on NLWs, the terminology not only refers to a wide range of weapons such as plastic bullets, chemical incapacitants and entanglements, but to anti-lethal and anti-material measures, communications and information technology and even air power.[4] The elasticity of the phrase, and thus the scope of technology that might be brought within discussions of NLWs, perhaps obscures the anti-personnel applications which are accredited with the most revolutionary potential. It is the anti-personnel aspects of weapons that will be focused on in this chapter.

More fundamentally though, there are reasons for doubting the viability of the category of non-lethality. Critics of non-lethal technology have argued their design and the manner of use often renders them lethal in practice. Aftergood[5] echoes the view of many in insisting that instead of portraying NLWs as an alternative to lethal force, they are better understood as complementing and thus enhancing such force. Attempts to present a benign public face to NLWs through relatively innocuous names

– such as rubber and plastic bullets, stun weapons, incapacitant sprays –
present a false picture.

Examples of the lethal consequences of non-lethal weapons abound.
Consider the case of plastic bullets in Northern Ireland. Campaigners point
to the 14 reported deaths and numerous serious injuries that have been
attributed directly to the use of these bullets. The guidelines of the Royal
Ulster Constabulary and British Army specify various conditions for the use
of plastic bullets: they should be deployed to disperse crowds, aimed at the
lower body, fired from a minimal distance, not fired indiscriminately and
not bounced off the ground. There is reason to suspect such guidelines have
been broken routinely.[6] In the case of the Gulf War, the actual deployment
of NLWs also raised difficulties for maintaining a distinction between lethal
and non-lethal. In one case, non-lethal minute carbon fibre spools were
used to cause short circuits in power plants and switching stations. The
resulting deaths from the lack of electric supplies cannot be understood as
non-lethal in character, nor can the use of conventional explosives after the
fibres to make sure the stations were rendered non-functional.[7] Given this
it is not surprising that alternative terms such as 'less-than-lethal', 'pre-
lethal' and 'soft-kill' have been advanced to designate non-lethal weapons.
Recognizing the problems of engaging in non-lethal conflict, organizations
such as the US Department of Defense have vacillated between preferring
non-lethal and less-than-lethal designations.

The difficulties in labelling are more basic than determining when a
weapon transforms from non-lethal to lethal or whether in practice it is
merely a complement to lethal force. While it may seem perfectly
reasonable to argue that a bullet from a M-16 or an AK-47 is more likely
to kill than plastic bullet, it is not clear what characterizations should be
attached to these objects. The notion of lethality has been called into
question as well. For instance, the International Committee of the Red
Cross[8] has studied the effects in wartime of conventional weapons and
found that wounds from fragmentation weapons and rifles kill roughly
20–25 per cent of the time. Likewise, with adequate hospital provisions,
anti-personnel mines rarely result in death.[9] In such a situation, ascribing
'lethal' capabilities to technology is not without its difficulties. From a
different perspective, the lethality of weapons such as flechettes, cluster
bombs, or napalm has been called into question by governments who wish
to refute humanitarian criticisms.[10]

Such observations are not just digressive points for reflection or
attempts to talk effects away; they have been at the heart of debates about
the control of weapon technology. The case of blinding laser weapons
illustrates how contrasting interpretations about lethality have
consequences for proliferation controls. Doswald-Beck[11] recounts
international discussions that built up to the banning of some forms of

laser weapons. Opponents to restrictions on weapons designed specifically to blind contended that any weapon that causes damage short of killing is more acceptable than conventional (read 'lethal') weapons. As there can be no fate for individuals worse than death, and lethal weapons are acceptable in war, then weapons with effects short of death should be acceptable also. The counter argument put forward to this line of thinking was that while individuals may think they are acting to kill, in practice, war rarely operates on such as basis (the 20–25 per cent death rate mentioned above). Blinding, however, results in largely permanent damage, which typically entails severe psychological and financial costs. Divergent interpretations about the acceptability of blinding lasers, in part, rested on what constituted the allowable range of military action to succeed in combat and what means could be used to achieve these ends. As expressed in various international humanitarian agreements, the methods and means of conducting war are not unlimited. Combatants should not be injured beyond that necessary to achieve a military purpose. In other words, the terms on which war is waged should not make simple assumptions about the lethality of warfare. Rather it is necessary to distinguish between the intention of individuals (e.g., to kill) and the likely results of their actions (e.g., to temporarily incapacitate) as well as legitimate and illegitimate methods.

The naming game surrounding the lethality of technology also enters into moral justifications for the trade in NLWs and limits on the responsibility for their ultimate deployment. The UK-based company Pains Wessex has exported tear gas to Kenya, where the improper use of such gas has been reported. In 1997, police in Nairobi broke up a pro-democracy rally in a cathedral by firing tear gas into the building. Tear gas is supposed to be released into open spaces for crowd dispersal rather than crowd punishment purposes. Responding to criticism from human rights groups over the possible implications of such sales and the need for stricter accountability in export controls, a company official is reported to have said, 'What do they want us to be transparent about? We don't make things that kill people.'[12] Here non-lethality serves as a means of diminishing the responsibility of supplier companies, and by implication the governments that approve such transfers, from the end use of the products.

Overall then, the lethality and non-lethality of weapons cannot be understood in a simple, self-evident fashion. Actors make strategic use of notions of lethality and non-lethality to advance particular images of technology. It is too simple to assume that humanitarian groups classify non-lethal weapons as lethal so as to justify strong regulation, while potential or actual users regard them as non-lethal and therefore acceptable (though this is often the case). The result is something of a twisted language that obscures some aspects while revealing others. As one commentator put it in relation to NLWs, 'The euphemisms and political correctness that surround the moral, legal, media and tactical aspects of warfare of the future are complex and bizarre'.[13] The contested status of NLWs complicates characterizing their consequences and thus formulating generalized guidelines for their control. No doubt as governments develop new ways of heating, disorientating, disabling, dazzling, incapacitating and calming, these issues about the characterization of the effects of non-lethal technology will become more pertinent.

Perspectives on Evaluating Non-Lethal Weapons

The question remains though, what can be said in the way of evaluating the effects of existing and future non-lethal weapons. This section overviews four approaches for assessing NLWs derived from concerns over international security, human rights, international humanitarian law and domestic policing. Investigating NLWs in this way risks conflating diverse situations where quite distinct laws and conditions apply. Such a wide ranging approach though offers the possibility of highlighting areas of overlap that might otherwise go unnoticed as well as common assumptions between diverse substantive areas. Despite the differences between the frameworks, each embodies more or less explicit notions of what constitutes unacceptable effects.

International Military Operations

Non-lethal weapons have come into vogue in military analyses because of their proposed affinity with the 'revolution in military affairs' and the changing nature of security. The rise in importance of counter-insurgency wars, border disputes, ethnic and religious violence, *coups d'etat*, national security and counter-revolutionary operations pose complex problems for US and European military forces. NLWs are claimed to provide an intermediate option between the use of lethal force and no force at all.[14] They provide a tool for quelling violent situations (perhaps involving non-military forces) or even pre-empting the outbreak of conflict. Drawing in the practices within police forces, those in the military must learn a continuum model of force where sliding force gradients are used depending on the particular demands of the situation.

Lovelace and Metz[15] give a wide-ranging analysis of the potential of non-lethal weapons in the context of (US) army operations. Here the utility of NLWs derives from the current global security environment and the need to maintain public perceptions of the benign intentions of military interventions. While making an overall positive case for continuing efforts on non-lethal weapons, the concept of non-lethality entails certain difficulties, including the possibility of making war more likely by making it less destructive, reducing the effectiveness of forces, risking soldiers' lives and blurring the rules of engagement regarding lethal force.[8]

Lovelace and Metz share with many other analysts an abstract concern for the fit between large-scale security changes and the potential of non-lethality as a concept.[4, 16] Given this, it is perhaps not surprising that the primary precautionary steps they note to prevent the inappropriate uses of NLWs relate to the establishment of proper military doctrines and rules of engagement. Operational concepts should determine the use of technology rather than the other way around. In relation to the issues of this article, the effects of NLWs depend on the terms of their operational use and of the guidelines established. The basic philosophy is one of ensuring that the proper rules govern non-lethal weapons. The lethality of weapons depends on the intentions and actions of users.

Lovelace and Metz also note the possibility of various unintended side-effects of NLWs. The variability of effects for individuals might have significant and unanticipated, perhaps even provocative or deadly, consequences. To prevent against such possibilities, it is advised the effects of technology should be clearly understood. 'Effects' here covers more than just physical effects. Lovelace and Metz advocate the:

> research and development of data bases to support the assessment of the psychological impact which certain actions are likely to have in certain cultures. This will help commanders to select the appropriate circumstance for the use of nonlethal weapons and guide the research and development community as they refine nonlethal weapons, thus linking operational imperatives and technology.[17]

Devising such a database would seem to require the ability to associate a particular technology and a certain effect (for a certain culture) with enough explanatory power to be of value.

In calling for a systematic understanding of effects, Lovelace and Metz share with many other analysts the desire to find an authoritative way of talking about the effects of non-lethal weapons. Despite the potential of analysis to resolve disputes, the decisiveness of testing is likely to be limited. The reasons for this are multiple and found in many other contested technologies: experts disagree on effects, values must ultimately come into play in any evaluations and effects vary across the population.[18] As suggested earlier, even if effects could be established beyond doubt, that would not necessarily lead to a particular course of action. The contested status of effects would no doubt complicate or undermine the suggestion by Lovelace and Metz that a database can be established that maps responses in a satisfactory way.

Amnesty International

Amnesty International has catalogued some of the threats to human rights posed by various technologies, including non-lethal weapons used for crowd control, security operations and incarceration.[19] Far from thinking of NLWs as means of ensuring benign interventions, Amnesty has documented the contribution of these weapons to repression. It has called for a halt on the export of technologies that might be used directly to violate human rights and the end to the export of equipment used for torture or the ill treatment of prisoners. In practice, in the past this meant calling for a halt to the supply of military, security and police technology which has the sole or primary purpose of contributing to human rights violations, which has been shown in practice to be used for human rights violations in a country, or which is sent to military, security and police forces responsible for previous human rights violations. Much of Amnesty's research and campaigning in this area centres around creating the conditions (e.g., transparency, accountability in export controls) that will prevent technologies, be they non-lethal or otherwise, from falling into the wrong hands.

This position is illustrated in a detailed study of technologies based on electro-shocks, including stun belts, tasers and electro-shock batons and shields.[20] Such electro-shock devices have a wide range of uses in prisons, riot control and torture. In contrast to past forms of electro-shock, modern shock weapons are both more severe and diverse. While electro-shock shields, taser guns and stun guns are all said to be safe if used 'properly', Amnesty maintains such claims are typically based on questionable research and require unrealistic operating conditions. Furthermore, the effects of such technology vary considerably depending on the context of use and the preparedness and physiology of those they are used on.

In order to prevent any security, police or military equipment from contributing to torture or degrading treatment, Amnesty advises a number of steps for governments and intergovernmental organizations. In short, the overall strategy suggested is one of prohibiting the spread of such technology, monitoring its use and assessing its effects. Such a strategy is not without difficulties, in addition to those associated with convincing governments of the merits of such measures. Operationalizing these guidelines requires determining which technologies are covered and which are unacceptable (e.g., those that are inherently degrading or those whose sole use is for torture). As in the case of blinding lasers and other weapons, disagreements like over the 'primary' and 'secondary' purposes of technology are not easily resolved.[21] Moreover, evaluations of the cruelty and degradingness of any technology are unlikely to be shared universally. Opening a space for some legitimate circumstances has often been a means for governments or companies to circumvent regulations on limits. With respect to trading, Western governments have traditionally distinguished between the legitimate (e.g., maintaining law and order) and illegitimate purposes (e.g., internal repression) of military and security equipment. While a transfer of the latter type might be called into question, the former type has often been deemed an acceptable basis for exporting equipment to regimes with questionable human rights records. Countries suspected of using electro-shock batons for torture, such as Turkey and Indonesia, have justified their need for electro-shock technology in terms of law enforcement requirements. Proving otherwise requires substantial effort.

Here a critique of particular technologies, rather than just the hands that use them or the ends to which they are put, would provide a basis for analysis that did not necessarily require documenting specific human rights violations. Such a basis would have the advantage of commenting on, as yet, undeployed technology. In limited cases Amnesty has claimed that the intrinsic design of a technology leads to the violation of human rights. In the cases of restraint electro-shock belts, for example, it has called for countries such as the United States to ban them altogether because they are cruel, inhuman and degrading.[22] While electro-shock devices attached to the body are inherently degrading, handheld electro-shock devices may be acceptable in some limited situations. However, unless the basis for these distinctions between technologies is clear and the inherentness of characteristics established, the applicability of such evaluations to future technological possibilities is limited (see below).

International Committee of the Red Cross

In an attempt to develop a systematic and forward-looking scheme for the assessment of 'non-lethal' and 'lethal' weapons, members of the International Committee of the Red Cross (ICRC) have offered guidelines for defining what constitutes an inherently abhorrent weapon.[23] International law prohibits the use of weapons that cause 'unnecessary suffering' or 'superfluous injury', yet operationalizing these principles into yardsticks for evaluation has been difficult. In March 1996 the ICRC organized a symposium on the effects of weapons in order to investigate the definition of weapons that result in superfluous injury;[9] out of that came the SIrUS Project.[24] In relation to these provisions, the SIrUS Project has offered 'clear and objective' criteria for determining what constitutes superfluous injury or unnecessary suffering. The criteria are based on the (explicit) assumption that health effects are the only valid basis for making such assessments. Health effects can be categorized into two sets of factors: use-dependent and design-dependent factors. So, a bullet may transfer a certain amount of energy to the body because of its design (design-dependent), but its effects also vary depending on its use (use-dependent).

The criteria specified are concerned with the foreseeable effects, which are design-dependent. Superfluous injury and unnecessary suffering are defined as the causing of:

- Specific disease, specific abnormal physiological state, specific abnormal psychological state, specific and permanent disability or disfigurement;
- Field mortality of more than 25 per cent or a hospital mortality of more than five per cent;
- Grade 3 wounds as measured by the Red Cross classification;
- Effects for which there is no well recognized and proven treatment.

These criteria are meant to apply to all types of weapons. Rather than regulating particular technologies on a case-by-case basis, the criteria cover categories of technologies based on their effects. Furthermore, they apply to both the means and methods of suffering; a weapon may be banned outright because its effects are 'inherent' within its design or it may only be banned in particular situations. The SIrUS report advises weapons dubbed 'non-lethal' should be matched against such criteria. Any weapon designed to blind, for instance, would fall foul of the first and fourth criteria.

Politics of Technology

In different ways the three approaches outlined above highlight particular aspects of non-lethal weapons and provide some guidance for their assessment. That said, in the main they are oriented toward the study of direct (more or less physical) effects. Non-lethal weapons thus enter into situations where they may be used or misused, but they do not fundamentally alter existing practices. While no doubt it is necessary to consider such direct effects, there are grounds for thinking an exclusive focus on this range of effects draws insufficient attention to certain aspects of NLWs. It is a contention of this chapter that if we want to understand the effects of NLWs, then we need to be sensitive to the subtle and complex effects they may have on the perceptions of its users, the organization of security forces, or power relations between operators and recipients. Although the indirect and wider implications of non-lethality are sometimes noted,[25] often this is done in a loose fashion. Let us now consider some attempts to speak about the varied and diverse possible effects of non-lethal weapons.

NLWs are sometimes grouped together with other technologies including multi-purpose riot tanks, prison technology and surveillance and data-veillance which are referred to as 'technologies of political control'.[26] One of the main messages from this literature is the need for an account of technology sophisticated enough to make the link between weapons deployment and institutional and structural conflicts in society.

Referring to non-lethal weapons as 'technologies of political control' casts a particular light on them that implies certain intentions behind their development and effects. Such a characteristic is just one type of a framing of technology. It has been argued elsewhere that it is possible to find diverse and conflicting assumptions about the relation between technology and society within accounts of military, security and police technologies.[27] Varied stories about the need, origin and justifications for technology underlie particular accounts. So, technologies can be thought of as mere tools that have good or bad uses, or instruments deliberately intended to express the (repressive) interests of the state. The development of technologies of political control is portrayed both as a highly rationalistic, efficiency driven process and one that is negotiated, politicized and uncertain. The assumptions made about the underlying relation between technology and society help organize our understanding of the important characteristics and consequences of NLWs, the knowledge that is important in assessing them and the strategies that could form a basis for their regulation. One of the most common assumptions is that technologies are neutral tools. Here, technology merely represents a means to an end. Non-lethal weapons are not intrinsically good or bad, rather it depends on the manner in which they are used.

In contrast to the neutral model, various studies suggest there is a subtle and complicated inter-relation of technology and action. Technology enters into and helps reconstitute existing patterns of social relations. Wright[28] has conducted a major study for the European Parliament on non-lethal weapons and other technologies of political control. Part of the argument of this document contends that the employment of NLWs since the 1960s in Northern Ireland might have fostered greater rather than lesser violence. This was due to the relations of dependence built up around NLWs and the inter-relation of lethal and non-lethal means of force. If an accurate portrayal, this situation cannot be understood merely by specifying the direct effects of technology. Rather, the meanings ascribed to technologies in particular circumstances bear on the manner in which these technologies are used.

Noting such observations about the 'politics' of technology, however, exposes some limitations of the approaches mentioned above. So, the ICRC assumes a distinction between design-dependent effects and use-dependent effects. Yet, it is questionable whether such a clear distinction should be made. As with the supposedly non-lethal properties of plastic bullets, the design of a technology may well 'encourage' its greater use or in some manner change the circumstances of its deployment. Neither of these points fit very well into the three perspectives outlined above, though no doubt they should contribute to the assessments made of NLWs.

In examining data-veillance and surveillance equipment, crowd control weapons, prison systems and interrogation and torture technologies,

TABLE 2

SUMMARY OF APPROACHES TO NON-LETHAL WEAPONS

	Lovelace and Metz (1998)	Amnesty International (1997)	ICRC (1997)	Wright (1998)
Topics of analysis	Military operations	Human rights violations; Horizontal proliferation	International humanitarian law	Domestic political control; Vertical and horizontal proliferation
Key aspects of NLWs	Proportional force response; Fostering public acceptability	Possibility of inflicting inhumane treatment	Severity of effects	Functioning of NLWs as technologies of political control
Primary relevant effects	Physical and psychological	Degrading, inhumane, violation of human rights	Medical and psychological	Political suppression
Critical points noted	Feasibility of mapping response patterns	Basis for attributing characteristics to technology	Separation of design and user effects	Characterisations of technological development
Recommendations	Data base of psychological effects; Rigorous rules of engagement	Enhancing transparency; Controlling trade; General reform	Stigmatise inappropriate technologies	Create accountability mechanism

Wright integrates the intended, direct and visible effects of non-lethal weapons with their less visible political implications. Such an achievement, however, is done by drawing on several opposing characteristics of technology.[28] So, they are both intended to achieve particular need-driven goals (e.g., improving the efficiency and effectiveness of forces) and their deployment is adrift in processes where few key decisions are made in systematic fashion.

The Affordances of Technology

Table 2 summarizes the approaches discussed in the last section. It notes the main topics of analysis, the key aspects highlighted, the main effects noted, recommendations advised and critical points noted for each of the four perspectives. The diversity of commentaries on non-lethal weapons attests to the scope for interpretation. The principal points of the last section include the complications of associating a technology with a specific effect, the related problems of establishing a definitive account of technology and the importance of perceptions of NLWs that individuals form. The previous section adds to the points raised earlier about the contested status of what 'lethal' and 'non-lethal' mean and the difficulty of ascribing properties to technologies in a straightforward manner. The problems involved in characterizing these technologies are just some instances of a much more general problem of how to justify particular readings of technology; that is, when does it make sense to close down the space for interpretations in order to advance a certain one.

Evaluating the category of electro-shock technology provides an illustration of the complications of justifying a particular reading. Opponents to the use of such equipment might argue that they are inherently unacceptable because of the character and unfamiliarity of the force involved. Yet, it would be difficult to sustain this interpretation in light of the varied uses of electro-shock technology and multiple situations in which it is deployed. Do, for instance, electrified stationary fences merit the same status as torturing devices such as electro-shock batons? Arguably not, yet both can kill and cause immense suffering. An electrified fence that previously guarded South Africa's border with Mozambique and Zimbabwe is claimed to have resulted in hundreds of deaths.[29] European companies are marketing electrified razor coil fences with kill or 'stun' capabilities.[30] As mentioned earlier, Amnesty International does not categorically reject all electro-shock technology, but makes a distinction between handheld electro-shock devices and those attached to the body. Presumably the criteria that inform any distinctions about acceptability of held hand and belt devices should, in turn, comment on the merits of related technologies such as anti-auto theft electro-shock devices recently approved for use in the UK.[31] Yet, if the criteria established are limited to direct physiological or psychological effects (e.g., whether they kill or merely 'stun'), then important questions about the inter-relation of technology and practices brought up in the last section are likely to be missed.

To summarize much of the preceding argumentation, the characterizations made of non-lethal weapons are hotly disputed. Debates on the merits of the NLWs involve much more than opposing descriptions of their direct effects. The interpretations of NLWs vary from tools that allow a minimization of force to objects designed to coercively reinforce existing social relations. Given the points made in this chapter, how should non-lethal weapons be evaluated? What sorts of conceptual and empirical repertoire might be helpful? On the basis of analysis is it possible to specify appropriate conditions, rules and monitoring procedures for NLWs? How might characteristics attributed to NLWs legitimize military or coercive intervention when these are not appropriate?

The purpose of the remainder of this chapter is not to formulate a systematic framework for evaluating non-lethal weapons that would answer each of these questions. Nor is it the purpose to specify where the distinction between lethal and non-lethal rests. The task here is narrower. The chapter will further elaborate on the key attributes of NLWs, which might provide a basis for their evaluation. In doing so it does not seek to resolve the ambiguity around what counts as a lethal weapon, but instead seeks to understand, at least partially, how that ambiguity could figure into an understanding of NLWs. The previous discussion contended that the specific characteristics of NLWs should be brought within an analysis of their effects. Merely elaborating the abstract fit between a category of weapons and particular situations of deployment is likely to gloss over any number of points. Rather, we need to consider the implications of the 'structure of the technology' for its use. Having said this though, much of this chapter has outlined the difficulties of establishing definitive accounts of the character of non-lethal weapons. The physicality of objects is not some readily identifiable ultimate constraint. Attempts to elucidate relevant properties of technology must be generalized enough to flag up generic issues, but also recognize that the meanings of these attributes are highly context specific and sometimes contested.

To provide a handle on some of the complexities of assessing non-lethal weapons, this section identifies potential key evaluative aspects. It situates a typology of the particular features of technology within a broader understanding of how those features ought to be treated. The intent is to provide an approach that is suggestive in assessing NLWs.

Table 3 lists a typology of the 'affordances' of non-lethal weapons, characteristics which might serve as a basis for assessing technologies. The term 'affordances' refers to perceived properties of an artefact that suggest how it might be used.[32] Any technology has a multiple number of perceived uses and effects. Speaking in terms of perceived properties is crucial as individuals sustain divergent interpretations of technology that are mobilized to justify particular courses of action. The properties of NLWs cannot be assumed as given since it is only through acts of interpretation and practice that actors form meanings about technology. Specific affordances might be drawn upon by those employing or critiquing NLWs to enable, justify or constrain some action. So, perceived assessments of the severity of effects believed to be afforded by different weapons (e.g., whether the effects are 'lethal'), comes to bear on how those weapons are deployed. Each affordance is meant to be generative rather than definitive in this sense of raising possibilities that need to be thought through for particular cases.

As the text below reveals, speaking in terms of affordances has a certain explanatory value. Because technology can serve alternative purposes from that which it was intended, the notion of affordances helps distance us from conceiving of the key features of technology as hard and fast characteristics that lead to definitive consequences. No single consequence derives from these affordances; rather it is necessary to consider how each may be realized. While from a pragmatic standpoint it may not be particularly fruitful to ask how meanings about the effects of anti-aircraft missiles or

TABLE 3
TYPOLOGY OF AFFORDANCES

Nature	
Severity	
Duration	
Indiscriminate vs. discriminate	
Effects of repeat vs. single use	related to medical and
Groups vs. individuals	psychological effects
Variability	
Delays in effects	
Visibility	
Signs of use	
Signs of using	
Reliability	related to perceptions
Gaugability	
Tuneability	
Ease of use	
Openness to reconfiguration	related to adaptability
Conditional use	
Specificity of use	
Availability	
Portability	related to mobility
Transferability	

tanks[33] are formed, this cannot be said for non-lethal weapons where so much of their promise rests on following 'proper' procedures and setting limits on their use to ensure non-lethality. The notion of affordances does not seek to avoid or somehow go beyond questions of politics or interpretation; rather these are seen as central to constituting the character of NLWs. In addition, framing a discussion of NLWs in terms of affordances helps avoid assuming the effects of technology follow on in a deterministic manner technology. Grint and Woolgar[34] argue that assumptions about the 'inherent' impacts of technology (e.g., the assumption that some deaths are 'inevitable' from the use of non-lethal weapons) can obscure a consideration of the reasons why some events take place. When people are persuaded that technology determines social relations, then justifying contingent outcomes becomes easier.

Affordances in Detail

By way of illustration let us briefly comment on each of the affordances and how they might affect our assessment of non-lethal weapons. The first grouping relates to the medical and psychological effects. Of all the groupings, this one is most often discussed in existing literature and to some extent has a standing in international and national laws. However, in terms of the arguments of this chapter, we need to go beyond trying to specify the effects that follow from perceived properties and instead then ask how each affordance is made sense of and how they may affect the use of NLWs. *Nature* includes the range of diseases, specific physiological or psychological states and disabilities that might result from NLWs. *Severity* relates to questions over the proportionality of effects and whether they are considered to inflicting unnecessary suffering. Note the nature, severity and *duration* (i.e., whether long- or short-term) form the basis for the SIrUS recommendations. Whether weapons are *indiscriminate* or *discriminate* or effect *groups* or *individuals* are somewhat inter-related concerns. With regard to the latter, rapid-response mobile fences and entanglement devices may trap groups rather than particular individuals. The results of non-lethal weapons are not always apparent, it is necessary to speak of the consequences from *single* and *repeated* use, the *variability* of effects across population and whether there are *delays*.

Many of these issues come into play in the example of CS incapacitant sprays. In an assessment of the use of CS incapacitants in English police forces,[35] officers reported significant time delays (i.e., more than five seconds) in the effects of the sprays in almost a third of cases. For reasons that are medically unclear, in ten per cent of cases the sprays had no perceived effect. The uncertainty thereby introduced obviously complicates a quick and clear appraisal of their effects and what subsequent action might be required. In general, the severity and nature of the effects caused to both operators and recipients of the sprays have been hotly contested.[36] Officers contemplating a certain use of the sprays do so with unconscious or conscious perceptions of what the sprays afford in terms of their effects.

Health effects, however, are just one set of affordances that need to be taken into account. The next grouping of affordances surround the perceptions of users, recipients and others regarding the sense of control, predictability and proportionality afforded by non-lethal weapons. The category of *visibility* issues includes the *signs of use* and *signs of using*, that is the 'residue marks' left by NLWs and the awareness that they are being utilized. With regard to the former, much of the perceived utility of electro-shock technology, especially for torture, stems from the lack of forensic evidence such technology is thought to leave behind.[37] With regard to the latter, in the case of blinding laser weapons, the ability to make lasers invisible and thus not give any chance for eye protection nor preparing for the effects was a significant grounds for criticizing this technology.[38] The 'technical' *reliability* of NLWs to perform as expected should not be taken for granted. For instance, the principal manufacturer of electro-shock stun belt technology to US police forces contends that their belts have only been activated 25 times, though nine of these were due to 'accidental' glitches.[39] *Gaugability* refers to the degree to which operators know how much energy they are applying. While police officers using batons might receive immediate feedback on the force they inflict, the same cannot be said of CS sprays. This is the case because of the physical separation of user and recipient as well as the time delays and variability of the effects. Electro-shock stun belts that can be activated from 300 feet represent a particular stark example of the distancing of subject and operator.[40] While the ability to gauge damage might not be highly relevant in the case of many conventional weapons, it is of considerable importance in ensuring the damage caused by non-lethal weapons is kept to a minimum. This feature is quite closely intertwined with the *tuning* of NLWs to different settings (e.g., 'stun' or 'kill'). This ability to tune gives raise to important questions about how individuals come to make decisions about what force to apply and in what fashion.

Technologies are more or less open to being adapted for various purposes and this is the topic of the third grouping. *Ease of use* means just how well a technology is suited for particular purposes. Some non-lethal weapons can be *reconfigured* in ways not often acknowledged. Security forces in Northern Ireland are alleged to have stiffened rubber bullets by inserting batteries into them, thereby increasing the damage caused.[41] Ensuring NLWs remain 'non-lethal' often requires adhering to certain *conditions* in the way they are handled. For instance, it may take a high degree of skill to ensure anti-materials technologies do not become anti-personnel; lasers not intended for blinding do not inadvertently blind; sticky foam guns do not entangle bystanders as well as their intended targets; etc. *Specificity of use* refers to the number of foreseeable purposes of technology (i.e., is the technology of single, dual, or multiple use). Here the question is whether a technology is fairly 'generic' and widely applicable (e.g., lasers) or highly specific (e.g., plastic bullets).

The last grouping of affordances deals with mobility of non-lethal weapons. The *availability* of technology is likely to bear on the extent and type of use. The widespread use of tear gas in Vietnam, originally introduced on grounds of minimizing civilian suffering, owes much to its easy availability in the war.[42] *Portability* refers to the transporting systems whereas *transferability* relates to the interdependence of a technology with supportive systems (e.g., riot control water cannon require frequent refilling).

There is a fair amount of overlap and fluidity between the items and the groupings in Table 3. The severity of effects, for instance, cannot be understood as independent of the conditions of the use of non-lethal weapons. This fluidity should not be a surprise, as any attempt to make generalized distinctions about such issues is unlikely to be tidy.

Outlining instances of such abstract affordances is not in itself enough for the assessment of non-lethal weapons. Rather we need to ask how these come to life in particular situations where wider social factors come into play, such as the accountability, monitoring and redress procedures in place. It might be worthwhile, for instance, to ask how those involved in the deployment of weapons construct an understanding of who are the targets of these technologies. The identity of 'likely' targets (e.g., related to their physiological state) ties into concerns about acceptability and conditions of use. In another way, the emotional distance of operators from recipients though mechanical and cultural means (e.g., because of racial or ethnic differences) acts to make violence more acceptable.[43] To reiterate a previous point, NLWs cannot be understood as possessing particular consequences that derive in a straightforward fashion from their physical properties. Characterizations of effects should be sensitive to changes in knowledge about effects, recognize the inevitability of some uncertainty and acknowledge the context-dependency of the meaning of non-lethal weapons.

In choosing to discuss the effects of non-lethal weapons in terms of affordances, a key concern is how these perceived properties are or are not realized. In which sort of situations are certain affordances likely to be of concern? What parts of an encounter involving NLWs are accorded capabilities, when and by whom? Perhaps most telling, who bears the burden for resolving the ambiguities surrounding NLWs? Such questions are important to address because notions about the intention, purpose and effects of this technology play a role in shaping its use. Not specific to weapons technology, Pfaffenberger[44] discusses technology development as a drama in order to emphasize the performative nature of statement and counter-statements made about the effects of technology. Such drama involves the creation of scenes (contexts) in which actors play out constructed and purposeful roles. By taking this sort of approach, properties of technology are not assumed as given, but rather it is necessary to ask in what situations particular behaviours and outcomes result. For instance, when do individuals distance themselves from blame and instead attribute technologies with intrinsic characteristics (e.g., being lethal)? The position advocated here is not merely one of seeing security forces changing in response to technology (or vice versa), but asking how police tactics, organization and technology co-evolve.

From these observations about the importance of framing the properties of technology in terms of affordances, it is possible to think a bit more systematically about the wider politics of technology. Authors such as Hinman[45] have alluded to the possibility of technology (in the case of electro-shock technology) to dramatically alter the conditions in which they are used. In this vein, the design and conditions of the deployment of plastic bullets may facilitate extensive use, calmative drugs may introduce possibilities for systematic rape and the mechanized use of chemical irritants in prisons may radically alter existing relations. In other words, the availability of non-lethal weapons may lead to more coercive acts. The basis for discussing such a possibility is rarely elaborated. However, as already stated, non-lethal weapons do not have political values embedded within them. It is not enough to project a technology into a context and expect a certain result, particularly consequences that are more far reaching in scope. Rather than imputing technologies with deterministic powers, it is necessary to think in a bit more detail about the source of effects. The affordances outlined in this section are a step in this direction.

The argument presented here can inform and further some of the frameworks discussed earlier. For instance, partially in recognition of the limits of past strategies, in 1999 the International Council of Amnesty International adopted new guidelines for the assessment of military, security and police technology. Besides opposing outright equipment whose sole or primary use is to commit human right violations, the international Amnesty movement will call for laws and regulations which:

- Suspend the use, manufacture, transfer and promotion of any type of equipment where credible evidence has shown that it may inherently lend itself to human rights abuse, pending the outcome of a rigorous, independent and impartial inquiry into the use and effects of that type of equipment.

- Prohibit the transfer and use of any type of equipment where credible evidence has shown that it may inherently lend itself to human rights abuse unless the receiving party has established rules (including mechanisms which enable the effective monitoring and observance of the rules) which regulate the eventual legitimate use of it and which are based upon international human rights and humanitarian law standards.

These points attempt to find a basis for critique which moves beyond considerations of who uses a particular technology or what are its primary functions; instead the focus is on the characteristics of technology and how those might be realized. The past evaluation of stun belts as inherently abusive rested on such a line of reasoning. Just how the case will be made that other equipment or weapons may inherently lend themselves to human rights abuse is something that will develop over time. Such an argument could draw on the type of criteria set out in the SIrUS report, but the affordances outlined here suggest an additional approach for substantiating the contribution of weapons to human right violations, one which draws attention to a range of issues often neglected. The points made about chemical irritant sprays in the UK, for instance, illustrate a different basis for critique.

Conclusion

This chapter has presented a dynamic process at work between the portrayals of non-lethal weapons and their effects. In practice, the notions of lethality and non-lethality are highly contested and there is fair degree of flexibility in the way technologies are characterized. Given the contested status of the terms 'lethal' and 'non-lethal' and the concepts of 'lethality' and 'non-lethality', this chapter has advanced the notion of affordances as a means of framing discussions about the effects of weapons. Rather than trying to set definitive accounts of the properties of non-lethal weapons, it has asked what features of these weapons might be relevant in suggesting how they might be used and how actors come to understand their effects. The discussion has stressed a myriad of aspects of technology that need to be taken into account if we are to achieve a viable understanding of their consequences, including actors' attitudes, conceptions of weapon capabilities, expectations about technological change and representations of NLWs. While assessments of NLWs should explore particular technologies in detail, this analysis has provided generic comments that might then help structure detailed accounts. Such an approach can usefully complement many existing, conventional analyses and can act as a basis for commenting about the possible implications of yet unrealized technology. Ultimately ensuring the appropriate use of non-lethal weapons requires making political choices that can not rest on analysis alone, but the analysis presented here should help clarify some of the important issues at stake.

Notes

1. NATO has adopted a policy on non-lethal weapons, see NATO. NATO Policy on Non-lethal Weapons. *NATO Press Statements* 13 Oct. 1999, http://www.nato.int/.
2. See Bunker R, ed. *Nonlethal Weapons: Terms and References*. INSS Occasional Paper 15, Colorado: USAF Academy, 1996; Lewer N, Schofield S. *Non-Lethal Weapons: A Fatal Attraction?* London: Zed Books, 1997.
3. In this chapter, the terms 'lethal' and 'non-lethal' are meant to refer to the character of the effects of weapons, whereas 'lethality' and 'non-lethality' refer to the wider strategic and operational concepts that guide the use of force and rely on weapons that deliver lethal or non-lethal effects.
4. See, for example, Bunker R, *Nonlethal Weapons: Terms and References*, 1996; Morris C, Morris J, Baines T. Weapons of Mass Protection: Nonlethality, Information Warfare, and Airpower in the Age of Chaos. *Airpower Journal* 1995; 9 (1): 15–29.
5. Aftergood S. The Soft-Kill Fallacy. *Bull Atom Sci* 1994; 50 (5): 40–45.
6. For an overview, see Committee on the Administration of Justice. *Plastic Bullets: A Briefing Paper*. Belfast: Committee on the Administration of Justice, 1998.
7. Lewer N, Schofield S. *Non-Lethal Weapons: A Fatal Attraction?* London: Zed Books, 1997.
8. International Committee of the Red Cross. *The Medical Profession and the Effects of Weapons: The Symposium*. Geneva: International Committee of the Red Cross, 1996.
9. Coupland R. 'Non-Lethal' Weapons: Precipitating a New Arms Race. *BMJ* 1997; 315: 72.
10. See Prokosch E. *The Technology of Killing*. London: Zed Books, 1995; Ch.6.
11. Doswald-Beck L. *Blinding Laser Weapons*. Papers in the Theory and Practice of Human Rights. No.14. Colchester: University of Essex, 1995.
12. Honigsbaum M. Arms Firm Avoids Export Ban. *Observer* 17 Jan. 1999.
13. Note 10: 72.
14. Becker J, Heal C. Less-Than-Lethal Force. *Jane's International Defence Review* 1996; Feb.: 62–4.
15. Lovelace D, Metz S. *Nonlethality and American Land Power: Strategic Context and Operational Concepts*. Carlisle, PA: US Army War College, 1998.
16. See, for example, Toffler A, Toffler H. *War and Anti-War*. New York: Little, Brown, 1994.
17. Note 8: vii.
18. Sarewitz D. *Frontiers of Illusion*. Philadelphia: Temple University Press, 1996.
19. Amnesty International. *Arming the Torturers: Electroshock Torture and the Spread of Stun Technology*. London: International Secretariat, 1997; Amnesty International, *Made in Britain – How the UK Makes Torture and Death its Business*. London: Amnesty International UK, 1997.
20. Amnesty International, *Arming the Torturers: Electroshock Torture and the Spread of Stun Technology*. London: International Secretariat, 1997: 1–49.
21. Note 11: 148–201.
22. Amnesty International. *Cruelty in Control? The Stun Belt and other Electro-Shock Equipment in Law Enforcement*. London: International Secretariat, 1999; Broger J. US prisons 'use electric shock belts for torture'. *Guardian* 9 June 1999: 15.
23. International Committee of the Red Cross. *The SIrUS Project*. Geneva: International Committee of the Red Cross Publications, 1997.
24. The SIrUS name is taken from the phrase 'superfluous injury and unnecessary suffering'.
25. Hinman L. Stunning Morality: The Moral Dimensions of Stun Belts. *Criminal Justice Ethics* 1998; Winter/Spring: 3–13.

74   THE FUTURE OF NON-LETHAL WEAPONS

26. Wright S. *An Appraisal of Technologies of Political Control*. Draft Report to the Scientific and Technological Options Assessment of the European Parliament. PE 166 499 Luxembourg 1998, www.jya.com/stoa-atpc.htm; British Society for Social Responsibility in Science, *TechnoCop: New Police Technologies*. London: Free Association, 1985.
27. Rappert B. Review Essay: Assessing Technologies of Political Control. *Journal of Peace Research* 1999; 36 (6): 741–51.
28. Wright S. *An Appraisal of Technologies of Political Control*: 1–100.
29. Monteiro T. 'Hundreds killed' by South Africa's border fence. *New Sci* Jan. 1990: 27.
30. Note 28: 55.
31. Shock awaits car thieves. *Express* 22 Nov. 1998: 7.
32. Norman D. *The Psychology of Everyday Things*. New York: Basic Books, 1988. Pfaffenberger B. Technological Dramas. *Science, Technology, and Human Values* 1992; 17 (3): 282–312.
33. This is not always the case. To give but one example, much of the effectiveness of the introduction of the tank into the First World War rested on unfounded beliefs within military and public circles about the capabilities of these machines. See TimeWatch, *The History of the Tank*. BBC2 TV Productions 3 Sept. 1998.
34. Grint K, Woolgar S. *The Machine at Work*. Oxford: Polity Press, 1997.
35. Police Research Group. *A Review of Police Trials of the CS Aerosol Incapacitant*. London: Police Research Group Publication, 1996.
36. Jenkins C. Is it safe? *Police Review* 1998; Nov.: 4. Trevisick S. *Dispatches: The Truth of CS*. London: Channel 4 and Liberty Publications, 1996.
37. Detailed medical examinations can often detect whether someone has been subjected to electro-shock technology, even after a few weeks. Nevertheless the forensic evidence is difficult to collect. Amnesty International tells the story of 'Roberto' from Zaire. In this case, Roberto 'was at first beaten with sticks, before an officer stopped the beating, saying *"it will leave scars and we will get complaints from Amnesty International."* The officer then ordered his men to use an electro-shock baton instead.' See Amnesty International, *Arming the Torturers: Electroshock Torture and the Spread of Stun Technology*. Executive Summary. London: International Secretariat, 1997.
38. Note 12: 8.
39. Note 26: 5.
40. Schulz W. Cruel and Unusual Punishment. *New York Times Review of Books*. 24 April 1997. http://www.nybooks.com:6900/nyrev .
41. Ackroyd C, Margolis K, Rosenhead J, Shallice T. *The Technology of Political Control*. London: Pluto Books, 1980.
42. Dando M. *A New Form of Warfare: The Rise of Non-Lethal Weapons*. London: Brassey's, 1996.
43. Grossman D. *On Killing*. London: Little, Brown, 1995.
44. Note 32, Pfaffenberger: 282.
45. Note 26: 11.

The Role of Sub-Lethal Weapons in Human Rights Abuse

STEVE WRIGHT

This chapter draws extensively on two reports commissioned from the Omega Foundation by the European Parliament.[1,2] The use of sub-lethal weapons in complex situations, such as the Northern Ireland and Israel-Palestinian conflicts, to augment rather than replace lethal weapons, raises a number of concerns. For example, in practical terms this means that these weapons can undermine, rather than keep, peace. Unfortunately, any radicalization of protestors induced by crowd control weapons is usually interpreted as a hardening of the conflict by ringleaders and results in the next, more severe, phase of the ongoing military logic being deployed with predictable results. Thus in Israel during the second *intifada*, following its use of plastic-coated steel bullets which kill, the army escalated riot control measures to helicopter gunships, rocket grenades and dum-dum bullets, all of which are prohibited by International Human Rights Law.

Many police and paramilitary forces are also using sub-lethal weapons to draw protestors into the firing line of more lethal systems or to capture dissidents for later processing – beatings, torture, disappearance and extra-judicial execution. (Examples from Amnesty International are presented below to illustrate these points.) Whilst America has specific problems with its citizens because of its constitutional policy of allowing citizens to bear arms, the Civil Liberties Committee of the European Parliament recently acknowledged that in this respect America has civilization but not as we know it. Consequently there are tremendous commercial and political pressures to deploy these weapons in one-to-one encounters, where they may indeed have a role in avoiding lethal consequences. However, as recent events at the World Trade Organization confrontation in Seattle have revealed, even in America it has been difficult for officers to use sub-lethal weapons for crowd control without going beyond the limits of the law.

Elsewhere, weapons such as electro-shock batons have become the universal tool of the torturer – yet serving officers continue to naively advocate electro-shock systems as exemplars of all that is most commendable in the ideal 'non-lethal' weapon. This chapter takes the view that not all policing is a social good and the proliferation of sub-lethal weapons is industrializing the scale of human rights abuse – but in a CNN media-friendly way. It is argued that the second generation of weapons pioneered by the US National Nuclear Laboratories (and elsewhere) will make this position much worse, that the research is flawed, that the cancellation of recent US projects (such as CLADS, acoustic weapons, bounding net mines and others) underlines the difficulties in making truly less-lethal weapons and says more about the nature and role of hype and public relations in the lucrative world of weapons procurement. As will be apparent in the discussion below, the European Parliament has been advised by Omega to be circumspect about such research (some of which has been shown in the courts to be corrupt) and to conduct its own social impact assessments of these technologies (Report PE 168.394/Fin ST2).

Whilst allegedly 'non-lethal' crowd control weapons have gained increasing prominence in recent years as tools for managing contemporary internal security demands, there has been a longstanding search for, and deployment of, such weapons throughout the twentieth century dating from their use in the former European colonies. Historic examples include so-called 'tear gas', wooden and rubber bullets, electric cattle prods and water cannon used by the British colonial forces in Cyprus and Hong Kong, who also developed a new set of riot control techniques.[3] New crowd control technologies encompassed not just the 'hardware' or apparatus of technical performance but also the 'software' – the standard operating procedures, routines, skills and associated tactics for deploying public control weapons. Thus these riot control tactics themselves can be considered as a technology capable of refinement and transfer and consisting of a spectrum of options containing increasing levels of coercion.

Proponents of such weapons present them as providing additional options for intervention between the use of lethal force and no response at all. A sliding scale of options has been presented which offers the possibility of defeating troublemakers with minimum aggression; less-lethal weapons allow force to be viewed as a continuum. Opponents have argued that this perspective is naive; the potential blurring of boundaries between lethal and non-lethal weapons and the associated blurring of boundaries between police and military operations has awesome implications for human rights, civil liberties and due-process and may actually undermine the effectiveness of state security forces.

There is still a criminal lack of imagination in understanding the human rights implications of transferring non-lethal weapons to the torturing states. Few nations have adequate export controls and most pay lip service to human rights implications without having any system in place to check real end-use. It has been suggested that legally binding measures that penalize companies who continue to knowingly export such weapons, when evidence exists to show they are being used in human rights violations, might radically change the situation. There is a great deal of room to put more legal teeth into this area and back moral responsibility with legal liability.

Currently Available Crowd Control Weapons and their Effects

The current market in crowd control weapons includes basic truncheons, side-handle batons, riot shields, kinetic impact weapons, rubber and PVC plastic baton rounds (including multi-shot riot guns), water cannon, which have been enhanced to fire 'slugs' or 'bullets' of water, marker dye and a range of chemical irritants to punish demonstrators, stun grenades, a wide variety of chemical irritant grenades, tear gas projectiles, aerosols and bulk sprayers (all based primarily on five disabling chemicals, CS, CN, CR, OC and Pava), electro-shock weapons including 50,000 volt riot shields and a range of hand-held shock batons with a capability varying from 5,000 to 300,000 volts. Increasing availability means that many more countries are now willing to actually use this technology. A key finding of this study is that at least 93 countries world-wide have deployed 'riot control weapons' including chemical irritants, kinetic weapons and water cannon (this is an underestimate because not all countries report on their crowd control arsenals). Of these countries, 43 also manufacture, supply or distribute riot weapons and ammunition. One of the most salient findings concerns the alleged effectiveness of these weapons as a humane substitute for lethal force.[2]

The present study found many examples, in 47 countries, of these so-called non-lethal alternatives being used in conjunction with lethal force and leading directly to injury and fatalities.[2] Again, this assessment probably underestimates the level of augmentation of lethal and non-lethal weapons deployment. A continuing survey conducted through Amnesty International reports that some states are ignorant about the crowd control weapon holdings of their military, security and police forces.

Harmless Weapons

Children are disproportionally affected by both rubber and plastic bullets. Doctors in Northern Ireland have reported chest injuries,[4] scalping, skin lacerations, fractures of limbs and facial bones, eye damage including eyelid laceration, damage to the eyeball or complete destruction of the eyes leading to blindness, damaged liver, ruptured spleen, damaged intestine and 17 cases of permanent disability or disfigurement caused by rubber bullets.[5] Plastic bullets were introduced to Northern Ireland in 1973 and had completely replaced rubber bullets by the end of 1975. Because they are more aerodynamically stable than rubber bullets they do not generally tumble in flight and therefore usually hit the target end on, delivering the maximum kinetic energy to a small area and creating the so-called 'target lesion'. They have proved to be more often fatal than rubber bullets, with impact to the chest or head particularly life threatening.[6]

Deaths and serious injuries have also been reported from the Occupied Territories in the Middle East where rubber-coated metal ammunition (balls or cylinders) and plastic/metal composite ammunition are used. The rubber ammunition is fired from an adapter fitted over the end of the muzzle of the rifle; the plastic ammunition is fired as normal from the rifle. One report cited more than 20 deaths caused by rubber and plastic ammunition. In 17 fatal cases, mostly in teenagers, death was caused by penetrating wounds to the head, lungs and heart or by non-penetrating blunt trauma to the head or spinal cord. The wounds were characteristic of high-velocity ammunition and missiles fired from low ranges.[7] Penetrating thoracic wounds were reported in 26 Palestinian casualties, two of whom died.[8] Damage to the eye caused by rubber or plastic ammunition was reported in 154 cases, of which 67 led to loss of the eye.[9]

According to a report by B'Tselem, at least 58 Palestinians were killed by rubber-coated steel bullets between January 1988 and November 1998. This figure includes 28 children under the age of 17 years, of whom 13 were under 13 years of age. They note that the figures are an underestimate because in many instances autopsies are not conducted and it is impossible to determine which type of bullet was involved.[10] The current *intifada* is already producing more evidence of the lethality of Israeli riot control measures.

Reports from South Africa detail severe injuries to the face and jaws resulting in soft and hard tissue damage, rupture of eyeballs, mandible fractures and severe bruising caused by rubber bullets.[11] A report by Dr Clifford Goldsmith of the South African Bishops' Conference described injuries including chest muscles 'ripped open' and other muscles 'ripped down to the bone'.

The health, medical and safety effects of stun weapons can be considered within two key areas, the intended design and use effects of the weapon and the effects resulting from the use (and abuse) of such weapons. A number of medical professionals and human rights organizations have highlighted the lack of independent medical, safety and scientific evaluation of stun technology.[12] Since the publication of these findings, a growing number of reports of injuries sustained from the use of stun weapons and reports of fatalities associated with the use of such weapons has made the need for such independent medical and scientific evaluation an urgent requirement.

According to the stated criteria, these weapons are designed to temporarily immobilize the recipient by delivering a series of short-duration, high-voltage pulses that lead to tetanic contraction of the body's muscles. Thus the body is involuntarily paralyzed as long as the current is flowing. Most manufacturers and suppliers of stun weapons state that the devices are medically safe and non-lethal (although some do caution against the possibility of fatalities). Indeed, most of the manufacturers' research and evaluation reports of stun weapons refer to non-lethality in relation to 'normally healthy people'. The impact on people with existing medical conditions has received less attention. For example, in December 1995 Harry Landis, a Texas correctional worker with a history of heart problems, was reported to have died after receiving two 45,000-volt shocks from an electrified riot shield.[13]

Injuries and deaths associated with stun weapons have been reported in Los Angeles[14] as well as other cities in the USA[15] and the UK.[16] Stun weapons have also been reported as having a causal link with the miscarriage of a pregnant woman.[17] The author of the report on miscarriage stated that 'As use of the TASER becomes more common, obstetrical clinicians may encounter complications from the TASER more often'. It has been reported that a woman killed her seven-month nephew with an electric stun gun in an effort to stop him crying.[18] Many suppliers of electro-shock batons have cited research undertaken in 1985 by a professor at Dusseldorf University, Germany. However, he emphasized that:

> his expert opinion only referred to an apparatus type 'Paralyzer' produced by the company Dicom Electronics Ltd, which was on the market at the time and that 'as far as I know this specific model we examined is not any more on the market ... it is the nature of things that risk assessments only apply to a very concrete version of an apparatus of this kind and that it is impossible to derive any general clearance certificates in respect to technical variants of these systems ... If manufacturers refer to my expert opinion as proof their products are 'unobjectionable', they do so without being authorized. [Translation of a letter to Dr Harold Hillman dated 10 May 1995.]

An increased awareness amongst the medical profession may have begun with articles reporting the possible involvement of stun guns in the sudden deaths of men restrained in the prone position by police officers.[19]

The Nobel prize-winning organization Pugwash has come to the conclusion that the term 'non-lethal' should be abandoned, not only because it covers a wide variety of different weapons but also because it can be dangerously misleading.

> In combat situations, 'sub-lethal' weapons are likely to be used in co-ordination with other weapons and could increase overall lethality. Weapons purportedly developed for conventional military or peacekeeping use are also likely to be used in civil wars or for oppression by brutal governments. Weapons developed for police

use may encourage the militarisation of police forces or be used for torture. If a generic term is needed, 'less-lethal' or 'pre-lethal' might be preferable.[20]

Abuse of Technologies

There are many ways in which either the design or the operational usage of crowd control weapons facilitates human rights violations. Abuse of these weapons consists of the breach of several layers of alleged safeguards. These include undermining set rules of engagement, a failure to ensure that any deployment of force is appropriate, transparent and accountable and the inherent characteristics of the technology itself that might predispose it to abuse. We also need to understand the context of and the extent to which police and military culture permit or even encourage misuse and whether or not these cultures punish members who breach known human and civil rights protocols. One of the most undermining trends in recent years is the militarization of the police – the cross-fertilization of what should be two very different operational cultures. This process is perhaps most pronounced in the United States where, for the last 20 years, Congress has encouraged the US military to supply new weapons and training to civilian police forces. This has institutionalized Special Weapons and Tactics (SWAT) teams in almost every state.[21]

Chemical Irritant Weapons

Chemical irritant weapons facilitate human rights abuses in several ways, including the infliction of street punishment, an activity explicitly prohibited in most guidelines. The UN Committee on the Prevention of Torture found a recent case of pepper gas being used by the Austrian police to carry out a racial attack and the US has a long history of such practices.[22] In the US there are also cases where officers have deliberately breached guidelines by using OC (pepper gas) to inflict street punishment. California police deputies pulled back the heads of environmental protestors, opened their eyelids and swabbed the burning liquid directly on to their eyeballs – an action which Amnesty International described as 'tantamount to torture'.[23]

Kinetic Weapons

Systematic misuse of truncheons has almost become a metaphor of the archetypal repressive police state. Modern kinetic energy weapons have also been systematically abused, particularly in Northern Ireland and Palestine, in a wide range of different ways. These abuses include doctoring projectiles to enhance lethality, breaching guidelines on use only as a last resort, firing below the minimum distance, firing at areas of the body that should not be targeted (such as head, face, neck or chest),[5, 11] shooting out of moving vehicles, use as street punishment during zone clearing operations, intimidation, denial of the right to peaceful protest, use in a sectarian or racist manner and use of disproportionate and excessive force.

Even when these weapons are used in a criminal way, the weapons leave no ballistic trail that could be used in an enquiry to trace the person responsible. A failure by the authorities to prosecute officers who use excessive force or who breach the guidelines has led to a culture of impunity, a disregard for the rules of law and has made the use of these weapons ordinary instead of extra-ordinary. This is especially true in Northern Ireland and Israel.[10] However, whilst failing to prosecute the officers, governments have acknowledged the misuse of the weapons by offering financial compensation to victims and families, often on condition of secret payment and the withdrawal of any criminal charges. Such 'cheque book litigation' has ensured that the full examination and public disclosure of the misuse of these weapons has been avoided by the authorities.

Electro-shock Weapons

Manufacturers and proponents cite the 'non-lethal' characteristics of such weapons as reasons to deploy them instead of lethal force or other types of crowd control weapons such as blunt trauma batons. These characteristics include the claims that stun weapons are 'non-lethal', do not cause blunt trauma injuries and leave no long-term physical effects. However, these characteristics are exactly those that have led human rights organizations and medical personnel helping to rehabilitate victims of torture to suggest that stun weapons have inherent characteristics that facilitate abuse, ill treatment and torture. What makes the misuse factor with stun weapons so high is not just the apparent degree of unreliability or technical variability and very narrow safety range of the technology, but also the practical difficulty of finding evidence which can prove they have been used to facilitate human rights violations.

Electro-shock weapons have been deliberately, and often repeatedly, applied to sensitive parts of prisoners' bodies, including their armpits, neck, face, chest, abdomen, the inside of their legs, the soles of their feet, inside their mouth and ears, on their genitals and inside their vagina, on their back and their rectum. Such practices are often combined with other forms of torture and ill treatment, including psychological torture, as described in this report.[1] In many cases electro-shock weapons are used against women in addition to rape or other sexual assaults. A Mexican woman, Layda Silva, reported how the Cobras security force had used electro-shock batons on her: 'I fell to the ground, but they carried on giving me the shocks – on my breasts, vagina, stomach, legs, all over my body'.[26]

In March 1997, Amnesty International published a report that documented electric shock torture and ill treatment in 50 countries worldwide since 1990. In 18 of these countries there was evidence that hand-held electro-shock weapons had been used to commit such human rights violations. These countries included Algeria, Austria, Bulgaria, China, Egypt, Greece, Lebanon, the Russian Federation, Saudi Arabia, South Africa, Sudan, Turkey, the United States of America, Uruguay, Vietnam, Yugoslavia (Kosovo) and Zaire.[12] Despite their stated adherence to the basic principles of international human rights laws, including ratifying international human rights treaties, governments continue to permit electric shock torture and ill treatment in prisons, detention centres and police stations.

Recent reports by the United Nations Commission on Human Rights document the use of torture in a wide range of countries including Turkey. One report stated that the 'torture of men, women and children continues to be widespread throughout Turkey, and people have "disappeared" or died in police custody'.[27] Other reports have identified specific police units that practice widespread torture including electric shock torture and ill treatment. Even in the United States, the record of electro-shock weapons is far from unblemished. Prison guards in Arizona, California, New Mexico and Texas have been accused of tormenting inmates with stun batons. There have also been allegations that officers of the INS (Immigration and Naturalization Service) have used stun weapons against detainees. The UN Special Rapporteur on Torture raised a series of cases where it had been reported that stun weapons had contributed to ill treatment or abuse.

Export of Crowd Control Weapons and Human Rights

Manufacturers or suppliers of crowd control weapons are found in at least ten of the 15 EU countries, Austria, Belgium, France, Germany, Italy, Netherlands, Portugal, Spain, Sweden and the United Kingdom. Of these, at least six (Belgium, France, Germany, Italy, Spain and the UK) have exported crowd control weapons to countries where human rights violations have been committed with such technologies (for instance, Bahrain, Egypt, Guatemala, Indonesia, Jordan, Kenya, Nigeria, Sri Lanka, Turkey, Zambia and Zimbabwe). However, effective parliamentary and public scrutiny of the trade and its impacts on human rights violations are made very difficult by the lack of comprehensive, timely and accurate data on transfers of the weapons at the international, EU and national level. For example, the voluntary UN Register of Conventional Arms Transfers, introduced in 1992, does not require states to provide details for most categories of crowd control weapon transfers. The following case studies provide further detail of the human rights problems associated with such transfers and the need for a common approach across all EU member states.

Kenya – Tear Gas, Plastic Bullets and Water Cannon

On 29 December 1997, in the run up to Kenya's elections, human rights non-government organizations (NGOs) raised concerns that the government's intimidation of opponents and violent disruption of political rallies threatened to undermine the polls. At least nine people were killed and hundreds injured when pro-democracy rallies were violently disrupted by security forces. The Kenyan student leader, Janai Robert Orina, described how 'Tear gas is a day-to-day experience for us ... There are times when the air around the city of Nairobi reeks of it'.[28]

On 8 July 1997, it was reported that Kenyan paramilitary police stormed the All Saints Anglican Cathedral, Nairobi, attacking pro-reform advocates who were sheltering inside. Reports state that police threw tear gas canisters inside the cathedral.[29] Following this incident, Amnesty International received the physical remains of tear gas canisters and plastic baton round canisters that had been used in Kenya. These canisters were identified as having been manufactured in the United Kingdom. The use of tear gas within confined spaces or when people cannot physically leave an area could be seen as a form of punishment rather than dispersal. A number of severe injuries and deaths have been attributed to the use of tear gas in such circumstances. Following campaigning by human rights organizations, in March 1998 the British government announced that since election on 1 May 1997 it had rejected £1.5 million worth of applications to export certain types of riot control equipment, including batons and tear gas, to the Kenyan police.

From the witness testimony and the physical remains of canisters, Amnesty International was able to identify that the manufacturer of the tear gas was a French-based company, Nobel Securite (formerly SNPE). Further press reports suggested that the water cannon had been shipped to Kenya from either South Africa, Israel or France. Even though the United Kingdom had refused export licences for tear gas and other riot control equipment, at least one French company had stepped in to fill the vacuum.

Indonesia

Over the last 20 years the international media and human rights organizations have documented numerous incidents where the Indonesian security forces have deployed both crowd control weapons and lethal force, often with severe consequences for peaceful protestors. However, with the exception of water cannon, what is striking is the lack of hard data on specific transfers of crowd control weapons.

The Indonesian security forces have deployed chemical irritants (CS), plastic baton rounds and water cannon that spray a mixture of water, chemical irritant and marker dye. Both German- and UK-manufactured water cannon were identified as being deployed on the streets of Bandung in July 1996. The German water cannon were identified as Mercedes Benz vehicles, but it is not known who actually constructed and supplied them. The UK water cannon were manufactured by Glover Webb (a subsidiary of GKN Defence) and supplied by Procurement Services International. Having permitted the export of three Tactica water cannon in 1994 and another six in 1995, the UK government agreed an export licence in December 1996 for a further seven water cannon and 303 Internal Security vehicles, despite numerous reports of the use of such vehicles in undermining human rights.

Despite the prolific use of crowd control weapons such as tear gas and plastic bullets,[5] there is little hard data on which countries are providing the transfers of such weapons and munitions. Past transfers of riot control weapons to Indonesia have included Mecar bullet-trap rifle grenades and small amounts of tear gas from the UK.[30] One possibility is that indigenous companies have established local production of such weapons and munitions through licensed production agreements. For example, PT Pindad manufactures a range of small arms and ammunition under licensed production agreements from European companies including FN Herstal, Browning (Belgium) and Beretta (Italy). In 1995 Jane's International Defence Directory reported that PT Pindad could supply a range of pyrotechnics including 'Grenades, anti-riot, tear gas CN, hand launched'. Such licensed production agreements (where a European-based company permits a third-country manufacturer to produce products under 'licence') raise grave concerns that European Union embargoes and human rights based export criteria will be undermined.

Limiting the Role of NLWs in Future Human Rights Violations

General Principles – Licensing

The manufacture, supply, distribution, brokerage and production of crowd control weapons should be licensed. All products should be subject to common criteria of quality control. In the event of malpractice or lax quality control, licenses should be withdrawn, production curtailed and legal sanctions imposed against those responsible. A publicly available harmonized coding system should be adopted across EU member states.

Past experience has shown that it is unwise to rely on manufacturers' unsubstantiated claims about the absence of hazards. In the US, companies making crowd control weapons have put their technical data in the public domain without loss of profitability. It would be good practice for all European companies making such weapons to be legally required to do likewise and for all research justifying the alleged harmless status of any less-lethal weapon to be published in the open scientific press before authorization and for any product license granted to be subject to such scrutiny. Legal force should also be given to the terms of engagement, which would make any officers who breached codes of conduct and guidelines for using crowd control weapons open to prosecution.

Electro-shock and Stun Weapons

Members of the European Parliament expressed surprise that the EU has actually given CE quality control markings for such weapons and foreign manufacturers such as those from Taiwan boast an official seal of approval in promoting their overseas sales (Taiwan bans such weapons for home use). Given that no EU state recognizes a legitimate use for these weapons, the STOA report recommended that this practice should be immediately terminated.[1]

Social Impact Assessment of Police Technologies

An alternative option would be to institutionalize the decision-making process so that common parameters are examined when deciding on innovations regarding crowd control weapons. The STOA report[1] asked the European Parliament to consider what might be involved in setting up the bureaucratic procedures to achieve this objective, along the lines of the current environmental impact assessment regimes. In practical terms, this would mean having a formal, independent 'Social Impact Assessment' of new police technologies before they are deployed. These assessments could establish objective criteria for assessing the biomedical effects of so-called less-lethal weapons undertaken independent from commercial or governmental research. Some of the other options covered in this article (such as health and safety and accountability of rules of engagement) might be appropriately used in this process to provide EU-wide recognized benchmarks.

Exports of Crowd Control Weapons to Human Rights Violators

EU member states have inconsistent policies in regard to controlling the export of certain crowd control technologies. If this continues, European companies and governments will continue colluding with human rights violations in states that have very poor human rights records. It would be hypocritical for the European Union to define 'areas of freedom, justice and security' inside its territories, whilst undermining the same rights of freedom, justice and security because of inappropriate and ineffective export controls on the supply, licensing and brokerage of crowd control weapons and munitions to other countries. There should be severe restrictions on the creation, deployment, use and export of weapons which cause inhumane treatment, superfluous injury or unnecessary suffering.

Using the same principled approach, effective limits should be set on the exports or licensed production of any crowd control technology, ancillary equipment and training, which is not seen as acceptable for use within the EU. Clearly, European states should not export crowd control weapons abroad that are deemed too hazardous for use on Europeans.

Notes

1. An Appraisal of the Technologies of Political Control (PE 166.49), Dec. 1997: http://jya.com/stoa-atpc.htm.
2. Crowd Control Technologies: An Assessment of Crowd Control Technology Options For The European Union (EP/1/IV/B/STOA/99/14/01): http://www.europarl.eu.int/dg4/stoa/en/publi/default.htm.
3. Ackroyd C, Margolis K, Shallice T. *The Technology of Political Control.* Harmondsworth: Pelican, 1977.
4. Shaw J. Pulmonary Contusion in Children due to Rubber Bullet Injuries. *BMJ* 1972; 4: 764–6.
5. Millar R, Rutherford WH, Johnstone S, Malhotra VJ. Injuries Caused by Rubber Bullets: A Report on 90 Patients. *Brit J Surg* 1975; 62: 480–86.
6. Ritchie AJ, Gibbons JRP. Life Threatening Injuries to the Chest Caused by Plastic Bullets. *BMJ* 1980; 301: 1027.
7. Hiss J, Hellman FN, Kahana T. Rubber and Plastic Ammunition Lethal Injuries: The Israeli Experience. *Med Sci Law* 1997; 37: 139.
8. Yellin A, Golan M, *et al.* Penetrating Thoracic Wounds Caused by Plastic Bullets. *J Cardiovasc Surg* 1992; 103: 381–5.
9. Jaouni ZM, O'Shea JG. Surgical Management of Ophthalmic Trauma due to the Palestinian Intifada. *Eye* 1997; 11: 392–7.
10. B'Tselem. *Death Foretold – Firing of 'Rubber' Bullets to Disperse Demonstrations in the Occupied Territories.* West Jerusalem: B'Tselem, 1998.
11. Cohen MA. Plastic Bullet Injuries to the Face and Jaws. *S A Med J* 1985; 68: 849–52.
12. Amnesty International. *Arming the Torturers: Electro-shock Torture and the Spread of Stun Technology.* New York: AI, 1997.
13. Amnesty International. *United States of America: Use of Electro-shock Stun Belts.* New York: AI, 1996.
14. Judge shocks noisy prisoner into silence. *Guardian*, 10 July 1998: 3.
15. Woman guilty in stun gun death. *Chicago Defender*, 1 July 1995. Section PG, Col.6: 1.
16. Burdett-Smith P. Stun Gun Injury. *J Accid Emerg Med* 1997; 14: 402–4.
17. Mehl LE. Electrical Injury from Tasering and Miscarriage. *Acta Obstet Gynaecol Scand* 1992; 71: 118–23.
18. A baby's stun-gun death. *New York Times*, 25 Nov. 1994.
19. O'Halloran RL, Lewman LV. *Amer J Forensic Med Path* 1993.
20. *Pugwash Newsletter*, Nov. 1997: 276.
21. Kraska P, Kappeler V. Militarizing American Police: The Rise and Normalization of Paramilitary Units. *Social Problems* 1997; 44: 24.
22. Hersh SM. Your Friendly Neighbourhood MACE. *New York Review of Books*, 27 March 1969: 41–4.
23. Amnesty International USA. Police Use of Pepper Spray is Tantamount to Torture. Press Release, 7 Nov. 1997.
24. Note 5: 480–86.
25. Note 11.
26. Weapons of torture. *Time Magazine*, 6 April 1998: 52–3.
27. Standing up for the victims? 1998 United Nations Commission on Human Rights. Amnesty International News Release, 12 March 1998. IOR 41/04/98.
28. Equipping Kenyan Repression. *Amnesty International UK Journal*, Jan./Feb. 1998: 14–15.
29. Seven die as police smash Kenyan protests. *Times*, 8 July 1997: 14.
30. Ezell E. *Small Arms Today.* Boston: Stackpole, 1988: 205.

Future Police Operations and Non-Lethal Weapons

JORMA JUSSILA

Incidents

Real-life examples illustrate how difficult it is to resolve conflicts without resorting to some use of force.

- A mentally disturbed man is wielding an axe and shouting threats. When a police patrol arrives he runs into the woods. The officers find him and he ignores all commands. When the officers release a police dog the man knocks the dog down with the axe and attacks the officers.

- An aggressive, possibly psychotic, person with no visible weapons shouts that he has HIV and that he will bite anyone coming near. He may be under the influence of narcotics and/or alcohol. The scene of the incident is a street with bystanders nearby. The person stands still for some time and then starts running away.

- A person armed with a knife enters the emergency ward of a hospital. He demands narcotics, saying that he has not had a shot in several days and that he has no money to buy the drug.

- A person armed with a pistol points the gun alternately at himself and the police shouting: 'Shoot me or I'll shoot you! I want to die!'

- A police patrol has detained a bodybuilder who is known to have used anabolic steroids for a long time. Due to amphetamine overdose he is in a psychotic state, raging wildly even when in hand and leg restraints. He is impervious to pain, dangerous to himself and others and does not respond to any communication.

- A dangerous, violent person is being transported from prison to a court of justice. He is known to be mentally unstable and unpredictable and also has hepatitis.

- A man has barricaded himself in the bedroom after seriously beating up his wife. The wife says the man has a gun and wants to commit suicide. When the police patrol step closer to the bedroom door the man fires a pistol at the officers.

- Outside a restaurant some white youngsters have attacked a young man belonging to an ethnic minority. They kick and beat him with clubs, threatening to kill him. With the last of his consciousness he pulls a knife, stabbing one of the attackers in self-defence and killing him. An angry mob has gathered around the unconscious, badly injured killer and continues beating him. They also prevent paramedics and police from approaching.

- A man has robbed a hotel cashier with a gun. Half an hour afterwards a police patrol stops a man on the street to check his identity. The man pulls a gun and commands both officers to lie face down on the street.

There are several people involved in a conflict: the object person; the police officer; his/her colleagues and often bystanders. It is the duty of the police to protect all of them by causing no more danger, damage or injury than is absolutely necessary and justifiable considering all aspects of the incident. Force is sometimes needed to save the object person's own life. Conflict resolution is not an easy task and the demands may be conflicting.

We have to face the fact that there are and always will be situations where lethal force is the only option and the outcome is fatal. Even if a low level of force is used an object person may die due to unforeseen circumstances, such as an illness of which the police has neither previous knowledge of nor any chance of detecting. There is no safe use of force.

Conflict resolution in a relaxed co-operative atmosphere is easy. Unfortunately police involvement means that something has got out of hand. Emotion and stress levels are probably high, impairing rational decision-making. Some object persons are rational, some excited and tense, some sober, others intoxicated or under the influence of narcotics, some are mentally ill, some are psychotic, perhaps due to mixed use of medication, drugs and alcohol, some are in a fanatic frenzy. It would be marvellous if we could communicate rationally with all of them. Alas, at some point some of these people cease to live in the same world as us. They do not recognize us, understand what we say or feel pain. They may even be fighting monsters that their chemical-soaked brains have created before them.

Police officers are often presented as well-balanced, friendly and intelligent mediators. A police officer's work is both stressful and dangerous and can take its toll. Combining it with bad management, peer pecking order contests and other pressures of life can turn even the best officer into a time bomb waiting to explode.

An institution like the police attracts people inclined towards controlling others and perhaps those trying to inflate their ego with the authority and uniform. In spite of sincere attempts to select recruits with great interpersonal skills we will still always have 'poor performers' because they either slip through the screening process or because life and experiences inside the police change them. Police officers are nothing but ordinary people with a badge and some special skills. They have the same faults as anyone and they commit similar crimes, albeit on a smaller scale.

Avoiding Desperation and Injury

To cure an illness it is necessary to remove its causes. Desperation, hate, hopelessness, bad attitudes, distorted beliefs, prejudices, low or excessively high self-esteem may cause both an ordinary citizen and a police officer to commit illegal and endangering acts which may result in injury or death. Injury may cause more desperation and hate. *Desperation Avoidance* may break this vicious circle.

All political and economic decisions in society and in any organization should be made with desperation avoidance as an objective. This is the kind of support decision-makers can give, not just to the local police, but to everybody because it helps in maintaining prospects of life and therefore a peaceful and safe community to live in. It is the kind of support police command can give to officers because it helps to maintain the officer's intellectual performance, which is the most crucial skill in a conflict. However, it must be recognized that even if social injustice could be totally removed we would still not have a society of entirely rational, sane, healthy, social and co-operating citizens.

Instead of 'non-lethality' we should talk about the concept and encompassing philosophy of *injury avoidance*. Ideal police operations should be effective without causing any short-term, long-term, temporary or permanent injury or death to anyone. Reality, however, is different. Injury avoidance is not achieved by relinquishing any justifiable use of force instruments, not even lethal means such as firearms, which are still necessary in some operational responses. Injury avoidance should be, and can be, applied to all use of force and to all weapon types. There are several ways to avoid unnecessary injury even when firearms are used.

Selecting Use of Force Equipment

Several judicial, medical, moral, social and tactical aspects must be considered before the use of a force instrument can be accepted in police use. Selection must always be based on research, be justified and be acceptable. Selection cannot be isolated from the context and tactics of policing. The use of force equipment must form a system whose components support and are compatible with each other.

The use of force is never soft, gentle and harmless. It always includes risks, the magnitude of which varies according to the situation and object person. Becoming a target for use of force means entering an exceptional and extreme situation where the body prepares for fight or flight. This can be fatal especially for a person with an illness, irrespective of what kind of force a police officer uses. A police officer cannot generally be expected to know or recognize the physiological condition of a person. A police officer's decisions can only be based on the average danger of various weapons and s/he can only do her/his best to avoid serious consequences. It is impossible to present any precise figures when estimating the dangerousness of weapons. Estimates can only be based on research, expert statements, probabilities and available statistics.

Judicial Frame of Reference

According to the principle of proportional right to self-defence, no greater force shall be used than what can be deemed defendable considering the danger of resistance, the identity of the resisting person and the need to maintain order. This principle of proportionality is well established in Finland, in accord with United Nations human rights principles. The force used and the harm caused must be in sensible proportion with the objective, situation and resistance. Therefore the force must be rated according to their relative injury potential, which can be considered as the product of severity of the injury, duration of the recovery and amount of pain caused. The instruments of force that police officers, routinely and voluntarily, test on themselves can therefore be considered the mildest.

Conflicts and danger vary and situations can change very rapidly (Table 1). It is therefore impossible to present a rigid deterministic pairing of justifiable force and certain threat.

When assessing the justifiability one has to look at the whole picture. The justifiability depends on the urgency and importance of the official task, the dangerousness of resistance and the available resources. It would therefore be justifiable for a solitary police officer having access only to basic equipment to use stronger means of force in a situation requiring immediate action than would be justifiable for a fully-equipped police unit. S/he would, however, have to judge whether the task at hand really required immediate action. The law takes into account not only what the situation was, but also how those involved perceived it. The use of force may therefore be acceptable if the police officer had justifiable reason to

TABLE 1
THREAT, BEHAVIOUR AND USE OF FORCE

Threat Level and Object Person Behaviour		Use of Force Instruments
Compliance	Able and willing to co-operate	Police officer presence

Passive Resistance	Talking has no effect, reluctant, passive and possibly resists verbally. May be in an incoherent state of mind.	1) Talking
		  a) persuasion
		  b) instructing
		  c) warning / challenging

Active Resistance	Active, but not assaultive resistance. Refuses to comply with instructions and orders without resorting to physical force, attempts to leave or escape, openly uses hostile language.
	
	Passive resistance creates an obstacle for an urgent police operation, significant property damage is imminent or persons are in danger.
		2) Unarmed methods
		  a) Pain control
		3) Armed methods
		  a) Restraints
		  b) Pain control
		  c) Electricity (taser)
		  d) Tear gas or irritant
		  e) Impact weapons
		  f) Dog
		  g) Medication
		  h) Lethal force

Physical Assault	Aggressive but unarmed or using only an impact weapon. Uses or clearly intends to use physical force or otherwise acts in such a way that according to common life experience injury to bystanders, police officer or object person him/herself is imminently possible. He/she for example kicks, strikes with fists or threatens to hit with some object.	
Armed Assault	Creates imminent threat of injury to others. Uses physical force or otherwise acts in such a way that according to common life experience injury to bystanders, police officer or object person him/herself is imminently possible. He/she for example points a gun, threatens with a bladed weapon, hypodermic needle or explosives.	
Lethal Assault	Creates imminent danger of serious injury or death. Uses physical force or otherwise acts in such a way that according to common life experience serious injury or death to bystanders, police officer or object person him/herself is imminently possible. He/she for example points or shoots a gun, threatens or attacks with a bladed weapon, hypodermic needle or explosives.	

believe that s/he was in immediate lethal danger. Like anybody else, a police officer has the right to defend her or himself (and others) against lethal threat with lethal force.

Looking at the use of force equipment in the light of the above scale, one can justifiably say that it is just as difficult to give up firearms as talking. Although no police force operates without firearms, there are excellent tactical solutions, like those used by Norway and the United Kingdom, where not all police officers carry guns.

Carrying a gun increases work safety, but only if the officer is proficient in handling it. If all officers are armed, there is a loaded gun present in every situation they face. On the other hand there is currently no other effective instrument of defence against a knife which can cause deep stab and large slash wounds with the risk of lethal loss of blood and severe injury to internal organs. Demanding an officer to use only a baton or an irritant spray in this situation would be inhumane and unreasonable.

It is often suggested that if police do not carry guns the criminals would feel no need to arm themselves and would not use guns against the police. The situation is not that simple. Most criminals arm themselves for protection against their own kind. It is a fallacy to believe that their actions are governed by rationality and that they would not use a gun to escape a desperate situation.

Social Frame of Reference

Police weapons must be researched, justified and acceptable to society. Their use must also be monitored by society. Almost anything can be used to interrogate, torture and punish. It is essential that the acceptability of a weapon type must not be confused with acceptability of methods. Misuse is an intention of persons, not a property inherent to a weapon.

Throughout world history weapons have been used for political control of dissatisfied citizens, terrorizing dissidents and torturing detained persons, to punish them, interrogate them or manipulate their opinions.[1] Resorting to this 'malevolent political control' indicates that loss of control in society has already occurred to some extent. A strong nation holds the characteristics of openness, justice and voluntary co-operation. Although threat of malevolent political control is remote in some countries, maintaining freedom and civilization needs vigilance.

The police require the confidence and approval of society in order to succeed. The best ways to gain these are open and accurate provision of information and credible control mechanisms. Although the use of force equipment is subject to open discussion, it is clear that the issue of police tactics cannot be entirely public. An ordinary citizen must be able to feel confident that every police officer using force has received appropriate and sufficient training and possesses the relevant, continuously updated skills and correct, responsible attitudes. Otherwise, liability issues can become both legally and morally difficult.

Police weaponry should not be indiscriminate.[2] Its effect should be accurately directed and with a high level of predictability. From this point of view, the traditional 'skip fire' wooden or plastic projectiles are unacceptable because of their poor accuracy. They are intended for shooting as ricochets from the surface of the street (hence 'skip fire'). To a large extent the trajectory of such a projectile is unpredictable and can hit a person's leg, genitals or face with equal probability, causing serious injury or in some cases death.

Cruel, inhumane and humiliating means of force are unacceptable. It is difficult to define what this means because general acceptability is often based on personal values, emotions, mental pictures and associations that can be, to some extent, manipulated. Many attempts are made to manipulate people's views on the use of force equipment. Unsubstantiated, emotionally appealing claims such as 'a hollow point expanding bullet tears a leg off' are often made. Some claims try to convince people that a certain weapon technology is like a virus contaminating its surroundings with malevolent behaviour.

Many technologies suffer from the stigma of misuse. It is said that baton ammunitions have been used against demonstrators with disastrous consequences and that electric weapons have been widely used for torture. People who make these claims ignore the fact that these instruments have also saved lives when correctly and legitimately used. They forget that high-level technology is not required for torture; an evil mind-set is more than enough. Misuse is not caused by the technology itself; it is caused by people, bad training, a lack of openness, lack of risk of being caught and non-existent controls.

It seems that public discussion is often based on colourful mental images and emotional beliefs rather than hard facts. Facts are less exciting and emotive. However, the danger is that bypassing facts may lead to decision-making based only on emotion and belief. We would not demand that a surgeon use a dull scalpel to carry out his job; likewise we can ill afford to equip the police with 'toothless' weaponry.

Police Tactics Frame of Reference

Readiness Delay

Readiness delay starts when the use of force is imminent or already justified and ends when the instrument of force is ready for use.

Deployment within the police organization determines the length of the delay. The further away an instrument is, either geographically or organizationally, the longer it takes until it is ready for use. However, it is impossible to place all instruments in all patrol cars and to train every officer to use them; just as it is impossible for a front-line officer carry all of equipment at all times.

Weaponry can be divided into categories: personal (continuously carried), support (patrol car placed) and special (brought to the scene of incident by a specially trained officer or an emergency response team). When planning the deployment scheme, one needs to take into account issues like geographical distances, number of officers available, population density and chain of command – that is, who decides upon the use of a certain weapon type. Deployment to reach optimum readiness delay can be a difficult problem to solve.

Tactical Range

Tactical range is defined as the maximum distance of accurate and effective application of a weapon. The tactical range of a baton is about one metre, irritant sprays and electric tasers work at two to three metres. The normal tactical range of a 9 x 19mm pistol is 10–15 metres. Accurate shooting with a pistol in a critical situation at ranges beyond that requires exceptional skill. The same can be said about a carbine of the same calibre; shooting with it beyond 65m becomes more difficult because the bullet trajectory deviates from the line of sight and therefore requires compensation. However, it must be noted that the maximum range of a spray or a taser is not much longer than its tactical range, whereas a pistol bullet can still be lethal when discharged hundreds of metres away.

Incidents justifying only short-range, less dangerous means of force are often very problematic and difficult to resolve without putting police officers in unreasonable danger. The combination of drug abuse and infectious diseases makes it unreasonable to always require close contact in order to subdue resistance. It is, therefore, virtually impossible to discontinue use of an irritant spray.

Response Time

Response time is defined as the time from the start of use until the desired effect has been reached. The effect is based on breaking the ability or will to fight. 'Breaking the ability' means disabling the object person's ability to fight irrespective of his/her will and state of mind. 'Breaking the will' means causing conditions like strong pain so that the object person does not want to continue fighting. It is obvious that the outcome of a will-based weapon is more difficult to predict, for example, a drug addict may be impervious to pain.

External Conditions

External conditions may exclude the use of certain weapon types. Climate conditions in Finland are exceptionally demanding. Poor performance in subzero temperatures has so far prevented the use of certain shotgun launched baton projectiles. It is also questionable whether some electric instruments will work reliably in wet and cold conditions and obviously tear gas cannot be used in hospital emergency wards. Even if some instruments cannot be used everywhere, they might be outstanding in certain situations. Therefore, general applicability cannot always be held as a criterion for selection.

Practicality

Anything that requires more than one person to carry or causes significant collateral damage or inconvenience to the officers is impractical. Even if such equipment is effective and available, in reality it may not be used at all – as with CS spray in Finland. Despite the size and efficiency the cross contamination and inconvenience to the arresting officers was so great that the spray was hardly ever used.

Cost Factors

Cost factors depend on deployment scheme. It is obvious that a personal weapon carried by each officer has a higher total cost than one used by a limited number of specially trained officers. The cost should not be considered as the determining factor and it should not prevent the purchase of best-quality equipment. Equipment needed to save lives must never be given a price tag.

The maintenance of a use of force instrument is a liability and work safety issue. A faulty instrument can be deadly for the target and police officer alike. All use of force equipment should, therefore, be subjected to periodic control to make sure it is fit for duty within the prescribed safety tolerances.

Technological Notes

Use of Force Instrument Comparison

No instrument of force is 100 per cent effective each time it is used. The response of a human being depends not only on physiological but also on psychological conditions. Therefore the only available comparisons of various weapons are generalizations. When a police organization considers adopting a certain type of instrument it must check medical databases and consult experts to make sure that all research reports and risks involved are known.

TABLE 2
FORCE INSTRUMENTS

Instrument	Effect	Based on	Range (metres)	Injuries when properly used	Selectivity when properly used	Recovery	Order	Notes
Capture nets, foams, goos	Involuntary	Restraint	3–10	Minimal	Reasonable	Quick	1	Impractical, very restricted applicability
Handcuffs	Involuntary	Restraint/pain	0	Minimal	High	Quick	1	
Taser	Involuntary, 0–1 seconds	Loss of muscle control	3	Minimal, dart wounds	High	About 1 min	2	
OC irritant	Conscious and partly involuntary, 2–5 seconds	Pain/involuntary closing of eyes	3–50	Slight	Reasonable, no significant permanent collateral damage to property	20–45 min	3	Dispersion properties different from those of CS
CS tear gas	Conscious, 2–10 seconds	Pain	3–50	Slight, but max. concentrations must be avoided	Reasonable, often permanent collateral damage to property	20–40 min	4	Some systematic effects on internal organs
Batons	Conscious	Pain	0–1	Bruises, contusions	High	Days	5	
Remote batons	Conscious	Pain	5–20	Bruises, contusions	High	Days	5	Firearms with less lethal ammunition
(Remote) medication	Involuntary, tens of seconds	Muscle relaxation or sleep	5–20	Hypodermic needle wound, skin laceration	High	Hours	5	Injection, 'tranquilising gun' or blowpipe. Presence of physician
Firearm with lethal ammunition	Conscious and involuntary, speed varies	Pain, loss of blood pressure, death	0–300	Serious	High	Injury may permanent	6	May be considered lethal when used as against property

Table 2 attempts to compare use of force instruments by some of their parameters. The effect column describes whether the effect is based on voluntary compliance due to for example pain, or is caused by involuntary disabling physiological reflexes. It is very difficult to estimate the time required for the voluntary effect to take place. Injuries are difficult to quantify. Reasonable is a very subjective term.

- *Capturing devices* – so far have not proven particularly useful. The nets get caught in everything between and to the side of the police and object person. Sticky foam requires a large heavy canister; the object person and his/her surroundings are contaminated with the foam, which is difficult to remove. How can the object person be transported after spraying?

- *Irritants and tear gases* – it is claimed that CS is safe and that there are no medical reports on OC but that OC is too strong. Comparing OC with CS is like comparing oranges with apples. It is claimed that OC is more powerful than CS, but on what scale? Studies available indicate that OC may be more effective and that is can be safely used in higher concentration than CS. There are no studies on the long-term effects of either but statistics, available studies[3] and the written statements of major Finnish medical and pharmacological institutions suggest OC to be safer. OC causes very little collateral damage compared to CS. Mixtures of CS and OC are said to be very effective although the Finnish police do not use them because no medical studies have been found on the effects of such combinations. OC has an inflammatory effect on mucous membranes. Does that boost the systemic effect of CS on liver and kidneys? If CS slightly increases the effectiveness, do we really need to increase the pain felt by an object person? However, do we really want to re-introduce the significant collateral damage caused by CS?

- *Electricity* – properly designed electrical equipment does not cause significant pain and has been statistically proven to be very safe.[4] These weapons are also very efficient, do not interfere with pacemakers or cause heart arrythmia. Written expert statements of major Finnish medical institutions support this view on the safety of properly constructed devices. However, the usefulness of pain inducing electric equipment for law enforcement purposes is questionable. Properly designed 'tasers' are energy transfer weapons. This is a class of weapon that has barely emerged and will have a future. Remote energy transfer weapons will no doubt replace the crude kinetic energy weapons we use today. Remote transfer of electricity by an ionizing laser has already been patented.

- *Medication* – it is occasionally claimed that it is inhumane for police to shoot a syringe filled with calmative or sleep inducing agent into an object person and that this would also represent an unauthorized use of medical authority. This is an exceptional instrument but it must be used under supervision of a medical expert. However, in some cases this can be the only alternative left to save the object person's own life.

- *Firearms* – devices for launching a projectile. There are many types of ammunition that meet the requirement of injury avoidance. The velocity of a properly designed and correctly used projectile is sufficient to cause a stunning blow and pain but nothing more serious. Irritant or tear gas filled projectiles can be used to deliver the substance at a distance. A very common misunderstanding is that hollow point expanding handgun bullets cause much more tissue damage than non-deforming full metal jacket bullets, in fact there is little difference between the two.[5] This is because the latter starts tumbling, causing an effect similar to that of an expanding bullet. Furthermore, some ordinary jacketed bullets cause excessive penetration and create unacceptable danger to any bystanders within the range of several hundred meters. Even when lethal ammunition is used it must meet some injury avoidance criteria.

Society, innocent bystanders, police officers, perpetrators – who do we want to protect the most? Analysis of non-lethal technologies, as outlined above, helps police make informed choices (both technical and moral) about which weapon is appropriate for a particular incident.

Notes

1. Wright S. *An Appraisal of Technologies of Political Control – A Report for the European Parliament.* Luxembourg: European Parliament Scientific and Technological Options Assessment (STOA), 1998: 166–499.
2. High Commissioner for Human Rights. *Human Rights and Law Enforcement – A Manual on Human Rights Training for the Police.* Professional Training Series No.5. Geneva: United Nations, 1997.
3. Vesaluoma M, Muller L, *et al.* Effects of Oleoresin Capsicum Pepper Spray on Human Corneal Morphology and Sensitivity. *Invest Opthalmol Vis Sci,* 2000; **41**: 2138–47.
4. Ordog GJ, Wasserberger J, *et al.* Electronic Gun (Taser) Injuries. *Ann Emerg Med* 1987; **16** (1): 73–8.
5. Jussila J. Wounding Potential of Handgun Bullets – The Relative Damage Index. *Annales Medicine Militaris Fenniae* 2000; **75**: 125–34.

Operationalizing Non-Lethality:
A Northern Ireland Perspective

COLIN BURROWS

Despite the current level of dissent republican terrorist actions and internecine loyalist violence, Northern Ireland is a society emerging from conflict. The last 33 years of what is euphemistically referred to as 'the troubles' have claimed the lives of 3,636 people in a province of just 1.5 million people. To place those figures in perspective, they would equate to a loss of life of 11,000 in Great Britain or 410,000 in the United States. The conflict has involved armed and highly organized and opposing terrorist groups. Terrorist action in Northern Ireland has claimed the lives of 302 police officers and 644 British soldiers. The last soldier killed was Lance Bombardier Restock; he was shot by a .50 calibre sniper rifle in 1997 whilst speaking to a motorist at a vehicle checkpoint. The last police officer murdered was Constable O'Reilly, killed as a result of an improvized grenade thrown during loyalist rioting in 1998. Of the 3,636 deaths that occurred during 'the troubles', 315 have been attributed to the military and 52 to the police.[1] Over the last 30 years the terrorist campaign has affected nearly every family and has resulted in violence and acts of terrorism and counter-terrorism in many countries throughout the world.[1]

The complex nature of the conflict has given Northern Ireland an international visibility, recognition and media attention out of all proportion to its size. Policing and military involvement in what has become known as 'operations other than war' and in 'peacekeeping operations' has become refined. What were once 'intensified counter insurgency operations' are gradually being modified to operational support of varying degrees of intensity, ranging from zero visibility to over-presence. This support is being provided to a police service on the eve of a transformation that is unprecedented in terms of scale and nature.

In contrast to the media image presented of Northern Ireland, the reality is in fact quite different. It is an area where normality and abnormality co-exist in a modern highly developed democracy within the United Kingdom. In terms of use of force, it is worth highlighting that Northern Ireland has only witnessed a total of two fatal shootings by police since 1992. This compares with an average of two per year over the same time period in the United Kingdom. This is not to underestimate the trauma of such events for all of the parties involved; each loss of life is a tragedy and the taking of life, sadly on occasion unavoidable, is the very antithesis of all that policing is about. Loss of life, irrespective of issues such as legality, creates a dissonance in belief and values of the police that goes to the very heart of all that they stand for. Unlike military personnel, police officers do not perform tours of duty in a distant 'theatre of operations'; the so-called theatre of operations is their home, the casualties of the conflict are their fellow countrymen and -women. The creation of an environment where all that work and live there can do so safely without the fear of violence reflects their chosen vocation.

Central to the philosophy developed within Northern Ireland is the concept that in all areas of law enforcement and security initiatives, the police are the lead agency and the military act in support of the police. This examination of how formative experience of violence has shaped and affected Northern Ireland society and policing approaches to responding to violence and managing conflict in Northern Ireland is within this context. The focus of this chapter is on managing conflict and responding to often potentially life-threatening violence within a public order context, but the lessons drawn and visions developed are more broadly applicable.

Responsibilities of the Police

The core responsibilities of the police have changed little since they were set out by the first commissioners of the London Metropolitan Police, Rowan and Mayne, in 1829. Whilst policing objectives have been re-cast in recent years under the title of the 'Statement of Common Purpose', they still reflect the basic principles of protection of life, property and the preservation of the peace. There is an expectation on behalf of governments and local communities that the police will be present at public gatherings and will, within the law, intervene to prevent harm and violence. Where necessary it is also required that police will be equipped and have a capacity to intervene and use force to contain and defuse violence.

Within modern democracies, there is also an expectation that police will operate within the legal system at all times and therefore collective and individual police action and or inaction will be subject to high levels of scrutiny. In post-incident inquiries on police use of force and in media review, issues such as preparedness, appropriateness, proportionality, necessity, over- and under-reaction will become key words in assessing effectiveness and acceptability.

In dealing with violent or potentially violent incidents it is essential that the police maintain an operational and evidential focus, thus bringing the situation under control and returning the community to a state of 'normality', a state that will vary from area to area depending on cultural norms. Responses must be measured in such a way as to be effective without contributing to the underlying cause of the conflict. It is therefore crucial for strategists and commanders to maximize operational capability and reduce tension and fear, not only within the community and communities with whom they interact, but also for their own officers. There is also a need to increase their command and operational resilience and the stress threshold at which their officers and communities can operate without reacting inappropriately to violence or provocation.

The use of force and policing are inextricably linked; compelling compliance and impeding resistance are essential elements of law enforcement. Clearly the use of force requires regulating and conditioning by the law and internal procedures and guidelines. Whilst strict legal tests of 'reasonableness' and/or 'necessity' set benchmarks that must not be breached, they do not represent the true test of legitimacy in the eyes of communities and opinion-formers. Issues of morality, ethics and personal and community perceptions are the determining factors by which individuals and communities determine the appropriateness of police action in any given circumstances.

'Appropriate', 'proportionate' and 'measured' are the tests that are applied outside the courtroom. If police operate within these perceived norms, they generally receive community support or at least acquiescence. Appropriate use of force assists in resolving conflict. Inappropriate or excessive use of force tends to alienate police from communities and perpetuate violence and the underlying causes of the conflict. It is therefore essential that those charged with planning and managing critical incidents, including public order events, have an intimate and real understanding of the underlying causes of the conflict and of the very real perception of the communities that they are charged with policing. Interventions must be designed to impact on the conflict and not perpetuate violence.

It is within this context that this chapter considers the issue of purpose effectiveness in relation to non-lethality within a public order context in Northern Ireland.

The Northern Ireland Context

Traditional causes of disorder elsewhere include sporting events, industrial disputes, social/political protest or local responses to a specific policing action, for example, a drugs raid or police shooting. In Northern Ireland, the causes of disorder lie in deep-seated divisions between two communities. The public disorder that erupts frequently involves naked sectarian hate and murderous intent.

Public disorder within the Northern Ireland context has traditionally involved three parties – riotous crowds from the two communities, either separately or in opposition, and the security forces. The disorder has been characterized by:

- The ability to mobilize large numbers of individuals at short notice;
- Organized large-scale events;
- Spontaneous outbreaks of violence, often in remote areas;
- Orchestrated province-wide rioting which stretches police resources;
- Co-ordinated attacks on security force personnel, commercial premises and the vulnerable homes of members of differing communities;
- The practised use of petrol bombs and other missiles against the security forces (often stock-piled);
- Infiltration of the crowd by terrorist factions who have used conventional firearms and grenades, both hand-thrown and weapon-launched.

The level and nature of violence sustained over the last 30 years is unequalled elsewhere in the western world. Police officers have stood steadfast between opposing factions maintaining and securing community interfaces, often at great risk to themselves. During serious public disorder, many police officers have received serious injuries and others, who have been cut-off from their colleagues, have been subjected to lethal attacks.

Protestors have become more organized and year-long preparations have resulted in the availability of an increasing range of weaponry of both lethal and less-lethal capabilities. These have included:

- High-intensity light beams;
- Lasers;
- Fireworks;
- Improvised grenades;
- Automatic gun fire.

The Royal Ulster Constabulary's formative experience of public disorder in the context of the current troubles was literally a 'baptism of fire'. The events in Belfast and Londonderry in 1968 quickly spread to towns and villages throughout the province. Police officers armed only with motorcycle helmets and 'bin lids' used as improvised shields had little protection from petrol bombs and images of officers set ablaze were burnt deeply into the consciousness of other officers throughout the 1970s. The vivid images carried on the front pages of newspapers and magazines demonstrated the vulnerability of officers to close quarter attack and the ineffectiveness of platoons of officers in built-up areas, especially those surrounded by high-rise flats. The events of 1968/69 also demonstrated the inability of traditional policing structures and approaches to public disorder to deal with mobilized communities that, for whatever reason, considered there was legitimacy in violent protest and revolt.

The Royal Ulster Constabulary (RUC), numbering less than 4,000 officers, albeit equipped with water cannon and CS grenades, were unable to contain the widespread public disorder that erupted. The Army, also equipped with CS and water cannon, were called in to support the civil power. Sectarian violence, often used as an opportunity by emerging terrorist elements to mount attacks with firearms and explosives, demonstrated the inadequacy of reliance on this type of public order equipment. The cumbersome nature of early water cannon and their vulnerability to attack showed them to be of limited value. In particular they were found to be difficult to manoeuvre, had limited utility due to water capacity and were very vulnerable to attack by firearms or explosives. Whilst the use of chemical agents such as CS as a crowd control agent in continental Europe is, in the main, non-controversial and culturally acceptable, the use of CS in 1968 Northern Ireland alienated law-abiding residents and business people. The image of 'gas masked' RUC officers and rioters still endures and is still graphically preserved on the street art that decorates gable walls in Londonderry.

The Use of Baton Rounds

The use of baton rounds has fluctuated considerably over the last 30 years. The early 'rubber bullets' used by the British Army were later replaced by 'plastic bullets'. Baton rounds became available from 1978 and were used by the police as well as the military. The peak usage of baton guns was during the civil unrest associated with the Republican prisoners' hunger strike of 1981; 29,601 baton rounds were fired and resulted in seven deaths. The levels of violence associated with the protests were unparalleled even by Northern Ireland's standards. During this period there were 1,205 protest rallies involving 355,000 people and ten prisoners died whilst on hunger strike. A total of 67 people were killed, including 15 soldiers and 15 police officers. Terrorist attacks included the use of automatic weapons and RPG 7 grenades. By the end of the year there had been 1,141 recorded shootings and 530 bombs planted, of which 398 exploded. In total 117 people died, including 21 police officers and 24 soldiers; republican activity resulted in 74 deaths, including those of ten hunger-strikers. Of the deaths, loyalists were responsible for 14, the Army for 14 and the police for three. Ability to deploy a baton round weapon system to maintain distance and restrain those intent on otherwise uncontainable violence undoubtedly mitigated against even higher death tolls.

Almost 15 years later, in 1996, a decision to prevent the annual Protestant Orange Order procession from Drumcree church along the predominately nationalist/Catholic Garvaghy Road resulted in a prolonged stand-off between loyalists and the security forces. The Drumcree stand-off became a *cause célèbre*. The threat of violence and the numbers involved in the protest became so large that the then Chief Constable, Sir Hugh Annesley, considered that the demonstration could not be prevented without resort to lethal force and decided that the march would be forced through the Garvaghy Road. In the province-wide rioting, mainly in nationalist areas outraged by the decision to force the parade down the Garvaghy Road, a total of 8,316 baton rounds were fired. There were no fatalities as a result of the use of baton rounds, but there were a number of serious injuries.

The Secretary of State for Northern Ireland's Review

Following the disturbances in 1996, the Secretary of State for Northern Ireland commissioned Her Majesty's Inspector of Constabulary (HMIC) to review the use of baton rounds in Northern Ireland. He concluded:

> from accounts and his own experience he has no doubt that the threats to the maintenance of law and order from some sections of society in recent decades require the law abiding to provide the police with adequate powers and protective equipment, and in Northern Ireland at this time this includes PBRs [plastic baton rounds].[2]

Whilst other police forces within the United Kingdom have experienced extremely serious disorder, it has not been of the sustained and lethal nature encountered in Northern Ireland. Nevertheless, HMIC, in the 1996 Inspection of the RUC, was deliberate in echoing the view of chief police officers in England and Wales in respect of a baton round requirement:

> It is the considered view of United Kingdom chief officers of police that the baton gun is an effective means by which rioters armed with petrol bombs and other lethal missiles can be kept at a distance, contained or dispersed. Equally, it provides a means of disabling at a safe distance those posing a serious threat to life that would otherwise require the intervention of officers at close quarters thus potentially placing them at great risk.[2]

A direct result of the HMIC report was the development of revised guidelines on the use of baton rounds throughout the United Kingdom. These guidelines were produced under the auspices of the Association of Chief Police Officers' of England, Wales and Northern Ireland. They were endorsed by the Home Secretary and the Secretary of State for Northern Ireland and issued in August 1999. A copy was placed in the library of the House of Commons. The guidelines provided that:

Where on the basis of a risk assessment of existing intelligence it is believed that serious rioting would involve a risk of loss of life, serious injury or substantial and serious damage to property, an officer of assistant chief constable rank may, with the prior agreement of the chief officer of police, deploy officers who are trained in the use of Baton Rounds as a less than lethal contingency in dealing with serious disorder.

Baton Rounds should only be used:

1. Where other methods of policing to restore or sustain public order have been tried and failed, or must from the nature of the circumstances be unlikely to succeed if tried – and –

2. Where their use is judged necessary to reduce a serious risk of:

   (a) loss of life or serious injury, or
   (b) substantial and serious damage to property where there is or is judged to be a sufficiently serious risk of loss of life or serious injury to justify their use.

   In assessing the risk of loss of life or serious injury occurring account should be taken of the risks to police officers and members of the emergency services as well as to members of the public and others.

These revised instructions tightened the circumstances when baton rounds could be issued and raised the authority levels required for firing, requiring the 'on the ground assessment of a designated senior officer' before a request for authority to use baton rounds would be considered. The guidelines also provided greater clarity in relation to the use of the weapon system by officers who were involved in the protection of property. The issue of the revised guidelines corresponded with significant enhancements in the personal protective equipment issued to police officers engaged in public order duties and with an overall reduction in the frequency and intensity of disorder in Northern Ireland. In 1999 there were a total of 112 baton rounds fired; in 2000 the number was reduced to 26.

Baton Round Development

There have been several distinct development phases:

- 1970 – Introduction of rubber bullets;
- 1973 – Introduction of plastic baton rounds (polyethylene and later polyurethane);
- 1983 – Introduction of 'rifled' barrelled weapons;

- 1994 – Introduction of a new design of baton gun;
- 1997 – Enhanced quality control measures;
- 1999 – Revised UK-wide police guidelines on use;
- 2001 – Introduction of round and sighting system.

In excess of 125,000 baton rounds have been fired in Northern Ireland since record-keeping commenced in the early 1970s. Between 1970 and 1989 there were 17 fatalities associated with the use of baton rounds. The loss of any life during public disorder is tragic and all police action is designed to minimize risk in so far as this is humanly possible. Where loss of life does occur, the impact on families, local communities and the officers involved is immeasurable.

The Heckler and Koch weapon system, introduced in 1994, significantly enhanced the ergonomics of the weapons system and enabled the weapon to be fired with enhanced accuracy. However, it is important to consider factually the lethality of the baton rounds used. Pre-1994, use amounted to one fatality per 6,500 rounds fired; since 1994 (after the introduction of the Heckler and Koch weapon system) 13,500 rounds have been fired and there have been no fatalities.

Independent Commission on Policing in Northern Ireland

The Independent Commission on Policing in Northern Ireland chaired by Chris Patten and established as part of the Belfast Agreement (Good Friday Agreement) of 10 April 1999 included a review of public order policing in Northern Ireland. In relation to the use of baton rounds the commissioners recorded that:

> All of us began our work wanting to recommend that [baton rounds] should be dispensed with straight away. But we do not want to see a situation in which the police would have no choice but to resort to live rounds sooner than would be the case today.[3]

The commission went on to recommend that an immediate and substantial investment be made in a research programme to find an acceptable, effective and less potentially lethal alternative to baton rounds. A research programme to that end has been embarked upon and a document reporting the outcome of its first phase and setting out the next steps in the work programme has been placed in the House of Commons Library.

Recent Development in Baton Round Design

In April 2001, the government announced as an interim measure, pending the outcome of the wider research programme, an enhanced baton round system designed to reduce the incidence of serious or life threatening injuries. The new baton round, designated the L21A1, has been issued to police forces in England, Wales and Northern Ireland and to the Army. The new round was deployed on an operational basis on 1 June 2001. Used, as it will be, with a new optical sight, the new baton round is more consistently accurate its predecessor and the probability of it causing serious or life-threatening injury has been reduced. Its injury potential was verified by an independent medical assessment, a copy of which was placed in the House of Commons Library by the Secretary of State for Defence. The statement announcing the introduction of the new round emphasized that there will be a smaller risk of serious injury or death, but that risk has not been eliminated and the new round, like the old, will be used in situations of public disorder only in accordance with the existing strict guidelines. Details of use in Northern Ireland will continue to be reported on every occasion and be copied to the office of the police ombudsman for Northern Ireland and to the new Policing Board, which was announced by the Secretary of State for Northern Ireland on 29 September 2001.

The statement went on to refer to potential use of the weapon system as a less-than-lethal weapon in non-public order situations by police officers in England and Wales:

> In addition to its possible use in situations of public disorder, the Association of Chief Police Officers (ACPO) considers that the improved accuracy of the new baton round makes it suitable for use in dealing with people who are posing an immediate threat to life in circumstances in which use of a firearm would otherwise be necessary.[4]

If the new baton round is used in this context it will be the first use of baton rounds outside Northern Ireland by British police officers. The new round is designed so as to ensure greater consistency in flight and to reduce the potential for unintended upper body strikes. This is intended to reduce the potential for serious or life-threatening injuries. A new optic sight will assist the firer in optimizing the accuracy of the weapon. The research and development costs have been in the region of £1.65 million and have resulted in a baton round system which features:

- Greatly enhanced accuracy and consistency;
- Flatter trajectory;
- Reduced bang and flash due to use of smokeless powder and new ignition system.

In comparison to the L5A6/7 batons, the new L21 round, used according to Ministry of Defence/Association of Chief Police Officers Guidance, is expected to:

- Be more accurate and therefore more operationally effective;
- Reduce the incidence of people other than the target being struck;
- Reduce the incidence of head impacts;
- Reduce the incidence of life-threatening injuries;
- Still result in very serious head injuries should it strike the head.

Promoting Non-Lethal Outcomes

The RUC employs the full range of conventional policing tactics, including a well-developed flexible and graduated response to disorder. This includes:

- Officers in conventional clothing;
- Officers in protective flame retardant suits, equipped with shields and batons;
- Vehicle tactics;
- Military support;
- Use of substantial crowd control obstacles.

The crowd control obstacles take the form of large steel container crates which the military place in strategic locations to block and deny access. Despite these elaborate defences and the use of well-protected officers supported by military colleagues, situations remain where, given the level of violence, the numbers involved and the threat of conventional and improvised weaponry, the need for distance necessitates that officers have available a means for maintaining distance and repelling attack from what would otherwise be overwhelming force.

Modern water cannon, as used in France and Belgium, are much larger than those previously deployed in Northern Ireland – approximate to the size of a large fire engine. Public order tactics used by the Belgian *gendarmerie* and particularly the use of water cannon have been evaluated and water cannon were deployed in Northern Ireland in both 1999 and 2000. Their use during loyalist rioting at Drumcree, Lurgan and Portadown proved effective and they have provided a graduated use of force option in dealing with certain levels of disorder. However, issues of manoeuvrability and water capacity still limit the situations in which they can be used.

Promoting non-lethal outcomes even in the face of life-threatening violence requires more than weapon systems. Officers have to be trained and equipped to operate with a high degree of spare capacity and confidence in the face of extreme violence. In terms of public order policing, this has included the provision of high-quality functional flame retardant suits to give officers the confidence to operate effectively, particularly when faced by petrol bomb threats. Officers are equipped with fire extinguishers and trained in role specific first aid skills. Protective helmets, properly designed and tested for the purpose (as opposed to modified motorcycle helmets), gloves, boots and appropriate heat-resistant underclothing to prevent heat transfer or burns are all on issue to officers deployed in public order situations. The investment in Personal Protective Equipment (PPE) is not only a Health and Safety requirement; it is an essential element in enhancing confidence, morale and resilience. The provision of such equipment reduces the fear and vulnerability experienced, but rarely articulated, by officers. Properly equipped and trained officers are more measured in their use of force and exercise higher levels of restraint than less well protected or trained officers.

The provision of appropriate personal protective equipment also expands the range of graduated flexible responses that can be utilized. Equally, it raises issues of image. To counter the Robocop appearance, and in particular any concept of anonymous unaccountable officers, all public officers in Northern Ireland have a personal identification number clearly displayed in large bold white numbers emblazoned on their protective helmets.

Promoting non-lethal outcomes also involves the provision of training for commanders, support personnel and units used in public order policing. Highly structured briefings, effective command and control arrangements and real-time community contacts which ensure effective negotiations with community leaders and influencers, even at the height of the disorder, provide for effective conflict management structures. Structured operational de-briefing is also an essential element of ensuring commitment and the development of a learning organization. Many of the improvements in communications, tactics and equipment that have been developed over the last five years have their genesis in the open and honest de-briefing organized after the traumatic events of Drumcree in 1996.

Conflict Management: An Alternative Approach

Tactics and equipment are but two elements of managing violence and responding to disorder. Developments in less-lethal weapon concepts continue to be kept under review. The government is vigorously pursuing a research programme to find an acceptable, effective and less potentially lethal alternative to baton rounds. Real progress is being made in managing conflict in often difficult and seemingly intractable situations.

Identification of stakeholders, including those within communities, who are reticent in direct dialogue with police, has been an essential element of managing conflict. The creation of fora, and where necessary third-party contact, for the facilitation of negotiation and mediation are high priorities. These have proved effective in making possible appropriate and timely interventions to manage confrontation. The development of real-time contacts before, during and after events is essential to the identification, resolution and avoidance of flash-point confrontations. A conflict management model must therefore be one that identifies the underlying causes of conflict and tension and agrees intervention strategies. It works on a variety of levels to de-escalate tension, avoid or resolve conflict and create a collective learning environment for all stakeholders.

The management of conflict is a core responsibility of the police service. Internationally, a great deal of effort has gone into the development of tactics, equipment and procedures for dealing with the actual encounter. Increasingly, attention is being focused on the skills and procedures required to diffuse critical situations, to negotiate acceptable outcomes and manage risk. Traditionally, police intervention in managing conflict and violence has been seen as being at the opposite end of the spectrum from the principles of community policing. To develop principles of community policing in isolation from those of managing conflict is to develop a two-tier dysfunctional approach to policing. A holistic approach is required if policing is to be developed so as to truly contribute to public safety and to enjoy the confidence of the community. Community partnerships, intelligence-led policing, the use of local officers as problem-solvers and the upholding of human rights are common to both the philosophy of policing with the community and to the management of conflict. Synergy between these two concepts is central to the development of an effective and acceptable policing service.

Conclusion

Mechanisms and strategies to avoid the creation of a downward reactive spiral to violence must be central to the planning of public order operations. Police and law-enforcement bodies cannot resolve the underlying conflicts in society; however, they can contribute to the management of conflict. It is essential that police officers involved in the planning of public order operations understand the underlying causes of the conflict, stand ready to intervene appropriately and are trained and equipped for their role. It is equally important that that planning and intervention strategies include those designed to diffuse conflict and promote non-violent outcomes to potentially violent encounters.

Notes

The observations and comments in the chapter are those of the author and do not necessarily represent the views of the Royal Ulster Constabulary.

1. McKitrick D, Kelters S, *et al*. *Lost Lives: The Stories of the Men, Women and Children who Died as a Result of the Northern Ireland Troubles*. Edinburgh and London: Mainstream, 2000.
2. Her Majesty's Inspector of Constabulary. *1996 Primary Inspection, Royal Ulster Constabulary*. London: HMSO, 1996: Appx.D, Para.21.
3. Independent Commission on Policing in Northern Ireland. *A New Beginning: Policing in Northern Ireland*. London: HMSO, 1999: Para.9.15.
4. Lord Bassam of Brighton. *Hansard,* House of Lords. 2 April 2001: Col.1591.

Non-Lethal Weapons Technologies: The Case for Independent Scientific Analysis

JÜRGEN ALTMANN

The NLWs Debate: Strong Claims, Little Hard Information

In the debate on non-lethal weapons (NLWs) which has gone on for about a decade, a wide variety of weapons principles were brought forward. Some NLWs, such as police batons, water cannon and tear gas, are well known and have been in wide use for a long time. The areas mentioned for 'new' NLWs are listed in Table 1. Note, however, that not all such NLWs concepts are actually new. Several had already been looked at (and many rejected for various reasons) decades before.[1-4]

It is questionable if psychological operations should count as 'weapons'. Computer-science weapons and operations represent a very complex field of their own and will not be discussed here. These technologies often come under the heading 'information warfare', which

TABLE 1
TECHNOLOGY AREAS FOR NEW NON-LETHAL WEAPONS

Mechanical	Physical Restraint – Person
	Physical Restraint – Vehicle, ship
	Kinetic Impact
	Acoustics
	Aerodynamics (vortex ring)
Chemical	Antipersonnel
	Antimaterial
Computer science	
Electromagnetic	Electric shock
	Electric short circuit
	Optical
Biological	Antipersonnel
	Antimaterial
Psychological	

comprises deception and perception management that are also aspects of psychological operations.

Fairly wide-ranging effects have been attributed to new types of NLWs, but little hard information is available. Many public proponents of NLWs have little or no scientific or engineering background. The little information that has been published often came as sensationalist articles in the military press, sometimes echoed by the general media. Because no alternative source of information was available, even critical scientists, when analyzing the NLWs movement and debate, came to rely on the manifestos of pro-NLWs groups or statements in journalistic articles where no detailed specifications were given.[5]

A few examples of claims made and reflected in scientists' articles include:[4, 6, 7]

- *Infrasound* – sound with less than 20-hertz frequency, causes disorientation, nausea and bowel spasms with effects from temporary discomfort to permanent damage and death. Against materials: embrittlement or fatigue of metals, thermal damage and de-lamination of composites are possible and, against buildings: even localized earthquakes.
- *Supercaustics* – chemical agents millions of times more caustic than hydrofluoric acid, can destroy optics or key weapons components.
- *Metal* – can be made brittle by means of an applied liquid metal.
- *Chemicals* – spread on runways or roads can dissolve rubber tyres or make them brittle.
- *Microbes* – can turn stored fuel into jelly and degrade explosives or even concrete and metal.

There are certainly NLWs technologies that are immediately credible, such as flash-blinding grenades, entanglement nets, electro-shock tasers or conductive ribbons to short-out open-air high-voltage power installations. Others seem at least plausible, such as polymerizing agents that clog engine air-intake filters. All such technologies may be problematic under criteria of international law, proliferation, human rights and the like and should be subject to thorough analyses in these fields.

On the other hand, the technologies listed above raise some doubts already with regard to their function, either principally or in an operational context. This chapter questions the technology and emphasizes the need to study it. Reliable knowledge of the properties of NLWs technologies is a precondition for their assessment with regard to aspects of military operations, international law, stability, arms race, policy and the like.

A Few Examples

To find out how credible are allegations about certain types of NLWs, a detailed study for each single technology is needed. Not many such studies exist at present; one is my own on acoustic weapons, referred to below. Even with the lack of detailed analyses, one can get some feeling for the validity of certain claims by using general knowledge from science, applying common sense and following the evolution of arguments over time. This section looks critically at a few specific claims. Some early statements will be compared with those in the book by J.B. Alexander that incorporates some of the early critical arguments against NLWs, stating, for instance, that there is no 'magic dust' for instantly putting people to sleep, that superacid-dissolution of tanks is impractical and that one should be cautious about releasing new organisms into the biosphere.[2]

Dissolution of Tyres

According to a brochure from Los Alamos National Laboratory, polymer treatment agents can be used to depolymerize certain elastomers (such as tyre components and rubber gaskets) and turn them into liquids.[8] The caption of an accompanying figure reads: 'Shown at right is an automotive tire immersed in a depolymerization agent'. The figure, on the other hand, shows a typical laboratory glass flask on which even the mark '100ml' can be read, so only a small piece of tyre can be immersed there. However, articles in the military press implicitly or explicitly produce the impression that the compounds 'could be sprayed onto runways and taxiways to crystallize and destroy aircraft tires'.[9] This concept raises questions about the volume of depolymerization agent needed to cover the $100,000m^2$ of a runway. If it is applied from close quarters, why not attack the runway or parked aircraft by traditional means? How thick a layer of tyre rubber will be affected by collecting, say, a few tenths of a millimetre of agent? How much of that will the tyre lose by rolling over a non-covered surface? Are there easy countermeasures, such as covering the runway with dirt and brushing it away again?

Rubber can of course be dissolved in a chemical, for example, sulphuric acid, if the rubber is immersed and the amount of chemical is sufficient. In Alexander's book, tyre attack by superacids (see also below) is discussed in terms of application by caltrops, rather than from the road, injecting the liquid into the tyre.[2] This is much more credible since it would keep most of the liquid in contact and would affect the rubber at depth. However, this suggests that the easy way – spraying the chemical on the asphalt – is unrealistic. As a better alternative, catalytic depolymerization is described which could be delivered in 'much smaller devices, but the effects would still be devastating'.[2] Unfortunately, exact information on the amount or the mode of delivery is not given.

Liquid-Metal Embrittlement

It has been claimed that liquid-metal embrittlement (LME) agents severely weaken almost any metal. Agents are generally clear and can be applied with a felt-tip pen, as a spray, splashed or brushed on.[10-12] Fast or slow action is possible which is 'especially important when LMEs are used against aircraft, ships, vehicles, railcars and railroad structures, bridge and building structural supports. Imagine how easy it would be to catastrophically damage a bridge's structural member using LME.'[12] Even more impressive is 'LME Graffiti', which is explained as:

> Graffiti used to mask an LME strike against a bridge or other target. Great potential for terrorist use. Example, phone call to law enforcement stating that an LME strike has been conducted against one of a number of bridges in a city using red LME graffiti.[13]

Could destroying a bridge really be as simple as drawing on it with a felt-tip pen? LME does indeed exist as a corrosive mechanism, but it occurs only under certain conditions. Only very few metals are liquid at normal temperatures, they corrode only certain solid metals and the process depends on factors such as grain size and metallurgical state. The basic mechanisms of this process are not yet fully understood.[14] Expert analysis is lacking, particularly concerning LME of steel at room temperature, but it is probably prudent to demand hard proof before accepting the claims as facts.

Superacid/Supercaustic Chemicals

Millions of times more caustic than hydrofluoric acid, these agents could be produced as binary compounds, in large amounts causing structural failure, 'damaging such things as asphalt surfaces, roof tops, or optical systems'.[6] Coupled with liquid metal embrittlement, supercaustics have 'tremendous potential' against 'weapons, vehicles, buildings and equipment'.[12] Alexander repeats such claims, stating that 'to destroy computer board circuitry requires a millilitre or less. To eat through a quarter-inch of aluminium takes about twenty minutes with a device the size of a soft drink can.'[2] Most probably, the aluminium has to be unpainted and the agent container has to be applied tightly, otherwise the liquid would leak. There are several unanswered critical questions, for example, what about steel? If one were able to apply such a container, why then not use a timed explosive charge? There may be special operations where silence is required, but the original claim appears very much weakened. Alexander strengthens this assessment by explicitly stating that 'dissolving tanks with superacid ... could be done' but 'the amount of superacid required makes it totally impractical'. Thus, we are left with 'critical nodes, such as optics, computers, or seals' by which larger weapons systems would become vulnerable.[2]

Microbes Turning Fuel to Jelly, Degrading Explosives or Hard Materials

A 1992 article states that 'under investigation are microbes that can turn large storage tanks of jet fuel into useless jelly'.[9] Alexander's statement, 'There is almost nothing that some microbe won't eat so the potential applications are extensive', was often quoted, by journalists as well as by some scientists.[6, 15] Statements like this invoke the notion that a small amount of microbe-containing liquid could be poured into a tank where the microbes would propagate, filling all its volume after a short time. Similarly, degradation of metals, asphalt or concrete was mentioned, producing the impression that one need just apply the microbes and the object would then be rendered useless.

However, microbes need specific conditions to live and propagate and for their degrading activity. The medium to be attacked may have to be available in small granules or even dissolved in aqueous solution. In addition, a certain temperatures, additional nutrients, and so on are required. Biodegradation of explosives, for example, has been studied almost exclusively in the context of contaminated soil. Even under favourable conditions, the process may take a long time, as is implicitly acknowledged in Alexander's book when he mentions microbial consumption of steel in the hull of the *Titanic*. On the other hand, he writes that:

> the degradation half-life, using P. fluorescens III on TNT is about one week. In other words, two weeks after inoculation, the explosives could explode with only 25 percent of the original power. One can imagine the impact of introducing these bacteria already at the point of manufacture. A very small number could have very large effects.[2]

Independent of the operational problems connected to such a scenario, it must be said that microbes do not attack solid explosive. First, to transfer the medium from crystalline form to solution needs a sufficient amount of water and second, nutrient minerals are required, as well as organic substrates, for the reduction of the nitro groups.[30] Thus, the degradation half-life cited above probably refers to TNT in solution.

Should future genetic engineering be able to design micro-organisms feeding only on fuel, solid explosive, concrete, or steel and doing so under operational conditions with sufficient efficiency (this is improbable but cannot be entirely excluded), it will have to solve the additional problem of restricting the effect to the intended targets and prevent uncontrolled spreading. A genetically modified microbial model system with increased degradative capabilities is being developed at the US Naval Research Laboratory. Additional modification is intended for self-limitation, 'either by timed "suicide" genes, or other alterations that prevent their persistence in the environment'.[16] One goal is to understand the biochemical degradation pathways in order to develop bio-mimetic chemical systems with similar effects; another is to devise defensive measures by which microbial attack on material could be frustrated in advance. Until such time, from a purely scientific and technical view, some research may be warranted, but not with the promise that these technologies are around the corner. From a legal and political view, it must be stressed that the Biological and Toxin Weapons Convention bans any hostile use of biological agents, that is, including use against material, as acknowledged by the US military.[17]

Infrasound

Inaudible sound with frequencies below 20 hertz has been said to incapacitate people by causing disorientation, nausea, vomiting and bowel spasms; the effect would cease when the source is switched off, without lingering damage.[10, 11, 15] A particularly colourful quote, assigned to 'a Pentagon briefing', is that acoustic weapons 'liquefy their bowels and reduce them to quivering diarrheic messes'. The same article reports that a United States defence contractor (SARA, Huntington Beach, CA) has built a device to make internal organs resonate, 'the effects can run from discomfort to damage or death'.[18] Alexander warns that 'care must be taken in the use of low-frequency sound so as to prevent permanent

TABLE 2

THRESHOLDS FOR ACOUSTIC EFFECTS AS STATED BY DEFENCE CONTRACTORS AND IN THE SCIENTIFIC LITERATURE

Defence contractors		Scientific literature[19, 20]
DASA 1995[21]		
Vertigo, headache, nausea, vomiting from infrasound	130dB	None up to 170dB
Eardrum rupture from infrasound	160dB	170–180dB
Eardrum rupture from blast	130dB	185 dB
SARA 1996[22]		
Intestinal pain, severe nausea from infrasound	110–130dB	none up to 170dB
Strong physical trauma, damage to tissues	140–150dB	only inner-ear receptors; eardrum rupture 160dB, only in audio range
Instantaneous blast-wave trauma, lethal effects	170dB	Lung rupture 200dB, death 210dB

DASA was tasked to make a detailed study on NLW for the German Ministry of Defence. The SARA numbers were given in a small-business innovative-research (SBIR) final report of 1996.

injury, or in extreme cases, death'. On the other hand, he states that persistent rumours that one French acoustician died from an infrasound blast 'as his internal organs turned to jelly' could not be confirmed by researchers.[2] However, in order to strengthen the credibility of anti-material use of acoustic weapons, Alexander relates stories which are hard to believe, such as the lifting of heavy stones by drums and trumpets in Tibet or the moving of concrete walls and weakening of railroad tracks by sound in the USA.[2]

To investigate the alleged effects of acoustic weapons, I have made a detailed study of sources, propagation and effects of strong sound which showed that most of the allegations were not true.[19, 20] For example, according to experiments documented in the scientific literature, infrasound did not produce disorientation, nausea, vomiting or uncontrolled defecation up to levels of 170dB – the maximum achievable in a special sealed infrasound chamber, which would be practically impossible to produce at some distance outside. (Note that with audible sound, ear pain starts at about 140dB.)[30]

In the low-audio region, on the other hand, intolerable sensations were produced above 150dB at frequencies between 50 and 100Hz – levels reached close to an aircraft jet engine, where a clear danger of permanent hearing damage exists.

Incorrect information is not only given in journalistic articles. In reports to their respective defence ministries, contractors used figures that contradict scientific reports (Table 2). Note that deviations of above 40dB occur, which means a factor of 100 in sound pressure.

A further indication of the incorrectness of the infrasound allegations is the fact that, after nearly ten years, the US Joint Non-Lethal Weapons Directorate (JNLWD) terminated its Non-Lethal Acoustics Weapons Programme in Autumn 1999 because the criteria 'showing a prototype device that could produce a reliable, repeatable bio-effect with sufficiently high infrasound amplitude at a minimum specified range' could not be met.[23] SARA is now developing a 'Multi-Sensory Distraction Device' combining visible light and repugnant smell with *audible* sound, maximally at the pain threshold.[24]

A common theme in these examples is that journalistic articles and the writings of NLWs proponents give incomplete information, which often seems to be based on hearsay. References are seldom given and therefore misunderstandings and exaggerations are likely consequences. Alexander has to be applauded for providing several pages of references in his book; this is unsystematic and scientific references are lacking but at least it indicates a starting point for follow-up research.[2] At the present stage, it cannot be concluded that all allegations mentioned are wrong because detailed studies are still lacking. In all probability, NLWs will turn out as another area where wonder weapons and magic bullets do not exist, as Alexander acknowledges, for antipersonnel sleep agents.[2]

NLWs Programmes in the United States

Some hints at the practical utility of NLWs technologies are also provided by the behaviour of the agencies responsible for new weapons and technologies. NLWs have to pass scrutiny before they are deployed. Such scrutiny is stricter for law enforcement than for the military. This holds especially true in democratic countries; people subject to legal force remain citizens with constitutional rights. Law enforcement institutions are subject to litigation and parliamentary oversight; the public media report abuses although of course the extent and the effect of the reports vary. The NLWs programme of the US National Institute of Justice (NIJ) provides a few examples (see Table 3). Sticky foam made its way into the major media world-wide – many people will remember the photo of a dummy covered

TABLE 3
NLW PROGRAMMES FUNDED BY THE SCIENCE AND TECHNOLOGY
DEPARTMENT OF THE US NATIONAL INSTITUTE OF JUSTICE

Public Acceptance of Police Technologies	1993	Laser Dazzler Assessment	1998
Aqueous Foam System	1994	Impact of OCSpray on Respiratory Function in the Sitting & Prone Maximal Restraint Positions	1998
Evaluation of Oleoresin Capsicum and Stun Device Effectiveness	1994	Evaluation of the Human Effects of the Sticky Shocker Topic	1998
Less-Than-Lethal Technology Assessment and Transfer Grant	1995	Evaluation of Vehicle Stopping Electromagnetic Prototype Device	1998
Law Enforcement Technology, Technology Transfer, Less-Than-Lethal Weapons Technology and Policy Liability Assessment	1996	Research and Establish a Computerized Database of Firearm Delivered Less Lethal Impact Munitions	1998
Law Enforcement Technology, Technology Transfer, Less-Than-Lethal Weapons Technology and Policy Assessment	1996	Biomechanical Assessment of Non-Lethal Weapons	1998
Ring Airfoil Projectile System for Less-Than-Lethal Application	1997	Preliminary Characterization and Safety Evaluation of Defense Technology's OC Powder	1999
Health Hazard Assessment for Kinetic Energy Impact Weapons	1997	Evaluation of Vehicle Stopping Electromagnetic Prototype Device – Phase III	1999
An Evaluation of Oleoresin Capsicum	1997	ROAD SENTRY Vehicle-Stopping PrototypeElectrostatic Discharge (ESD) Device	1999
Pepper Spray Projectile/Disperser	1997	Applicability of Non-Lethal Weapons Technology in School	1999

The starting year is from the grant number; many programmes are already finished. Data sourced by searching http://nij.ncjrs.org/ portfolio for 'Less-Than-Lethal Technology' (26 April 2001).

with glue.[5] This worked well, in fact too well. Cleaning it up took so much time that the NIJ found it 'hopelessly impractical for law enforcement use'.[1] A different, but also negative, verdict was passed upon anaesthetic agents: knocking out a hostage-taker fast enough was simply not compatible with a reasonable safety margin with respect to the hostages.[1]

Constitutional and compensation arguments do not hold for armed forces operating outside their home country.[1] However, because of the declared goal of NLWs, 'minimizing fatalities, permanent injury to personnel, and undesired damage to property and the environment',[25] military NLWs agencies work under an obligation to take the situation of victims into consideration. The 'CNN factor' provides an additional motive. Military agencies will in any case put high emphasis on proof of effectiveness under realistic conditions. The JNLWD of the US armed services has already terminated several programmes. In addition to

TABLE 4

PROGRAMMES OF THE US JOINT NON-LETHAL WEAPONS DIRECTORATE

Acquisition Programmes	Concept Exploration Programmes
40mm Crowd Dispersal Cartridge	Non-Lethal Slippery Foam
66mm Vehicle Launched Non-Lethal Grenade	Area Denial to Personnel
Bounding Non-Lethal Munition*	Clear Facilities of Personnel
Canister Launched Area Denial System*	
Modular Crowd Control Munition	**Pre-Milestone 0 Programmes**
Non-Lethal Rigid Foam	Active Denial Technology
Portable Vehicle Arresting Barrier	Ground Vehicle Stopper
	Unmanned Aerial Vehicle – Non-Lethal Payload
Joint Integration Programme (some of the equipment)	Vessel Stopper System
Hand thrown or shotgun launched stingball grenade	
Stun grenade	**Technology Investment Programmes**
Caltrops	Non-Lethal 81mm Mortar
Riot control training suit	Frangible Mortar
Body shield	Airborne Tactical Laser
OC (pepper) spray	Pulsed Energy Projectile
Search light	Odorous Substances
Shotgun with launching cup	Microencapsulation
12 gauge and 40mm non-lethal munitions	Bio-Materials Survey
5.56 rifle muzzle launched ordnance	Overhead Liquid Dispersal System
Ballistic face shield	Taser Landmine
Shin guards	NLW Guided Projectile
Riot batons	

Programmes marked with an asterisk have been terminated. The acoustic-weapons programme, terminated in 1999, is no longer listed.
http://www.iis.marcorsyscom.usmc.mil/jnlwd/Programs/Acquisition.htm,/JIP.htm,/CEP.htm,/Pre-Milestone_0.htm,/Tech_Invest.htm (8 Jan. 2001).

ending the Non-Lethal Acoustics Weapons Programme mentioned above, the Bounding Non-Lethal Munition and the Canister Launched Area Denial System were recommended for termination due to cost, schedule and performance issues. The general impression about JNLWD programmes is that they consist of simple, non-speculative technologies (Table 4). One reason for this may be that the primary focus is on requirements arising from continuing operations (for example, in Kosovo).

It will be interesting to watch whether and how NLWs agencies will deal with the more exotic technologies for which wide-ranging claims have been made by their proponents.

Dangers of Absent or Incorrect Assessments

Whenever decisions are taken on the basis of wrong or incomplete information, dangers arise. The history of the NLWs debate illustrates some of these. Claims were made by proponents without giving valid references. Journalists reported what they had heard (or understood) of military projects. Instead of demanding evidence, later authors took the assertions for granted. As a consequence, studies from military academies as well as articles and books from peace researchers repeated this information, mutually increasing apparent credibility. Many of the early claims are contained in a 1997 compilation of NLWs terms and references and were repeated in 2000.[13, 26] Despite this criticism, the 1997 report is valuable in its provision of some 640 references about the NLWs debate. However, it contains only very few scientific articles. It is thus no wonder that international-law studies discuss them as if they were reality,[27] or that Parliamentary Committees echo them.[28]

Advertisement of NLWs as too highly efficient with too few disadvantages and too quickly available may lead to unnecessary waste of taxpayers' money. This, however, is a secondary problem. Other dangers based on unrealistic expectations about technologies – changes of military strategy and faster political decision for military intervention – are more important. Arms control treaties can come under criticism, for example, the Biological Weapons Convention because it forbids anti-material use of microbes. In general, there is the danger that by setting unrealistically high promises for NLWs, decision-makers may rely on the technological fix instead of doing the tedious work needed for political solutions.

There are dangers too, for peace researchers, international law experts and NLWs critics; if they warn of some unfounded weapon effects, their arguments will be unsustainable in the long-term and may decrease their credibility. By focusing on threats that are not really important, they may reduce their attention to more dangerous technologies.

Preventive Arms Control

Many types of NLWs have not previously been deployed with the armed forces or law enforcement agencies. The latter are – in democratic societies – subject to control by civilian decision-making bodies, the law and the public at large. Procedures exist which can ensure that, in the introduction of new weapons, the rights and interests of victims as well as bystanders are considered adequately. For armed forces, on the other hand, systematic judgement on new weapons exists only in embryonic form, that is, they are considered under the laws governing warfare. Article 36 of Additional Protocol I to the 1949 Geneva Convention obliges States Parties to determine whether a new weapon, means or method of warfare would be prohibited by international law. There may be more arguments why certain new weapon types can have negative consequences for peace and international security. Limits on new weapons need to be agreed upon internationally, with verification, to be effective. The concept of *preventive arms control* exists as a systematic approach to this problem.[29] Because NLWs are an important example in this field, the concept is briefly explained here.

Preventive arms control is a form of qualitative arms control that limits militarily relevant technologies before the corresponding weapons are deployed, in other words, at the stages of research, development and/or testing. It aims at avoiding many of the problems highlighted by the cold war, that is, the introduction of new weapons types (such as the hydrogen bomb, intercontinental ballistic missiles and multiple independent re-entry vehicles) with the consequences of reduced warning time and destabilization and only later, after a tedious negotiating process, agreement about limits which were usually much too high.

Preventive arms control is a four-step process. (For law-enforcement, the steps can be applied in modified form.)

- *Prospective scientific analysis* – For each technology and each operational scenario, the relevant facts are to be described. Aspects to consider include technical properties of the weapon, propagation of the effect (range, selectivity, etc.) and effects on the target.

- *Analysis of military-operational aspects* – This includes the likely forms of use against various targets but also unusual applications or accidental effects. Also, the question of countermeasures (and of counter-countermeasures) has to be studied.

- *Criteria of preventive arms control* – The technical and operational facts or probabilities have to be assessed. The most important of these are:
  - Dangers for arms-control treaties;

- Destabilization;
- Arms races;
- Proliferation;
- Concern for the international laws of warfare;
- Dangers for the environment;
- Risks for civilian society and human rights.

In the case of NLWs, assessment is needed for each operational scenario: armed conflict, peace-enforcing operation or peacekeeping operation. In law enforcement scenarios, use against crowds and use against single criminals (in varying circumstances, from jail revolt to hostage taking) has to be considered; here, mostly the latter criteria apply.

- When a criterion gives reason for concern, possible *limits*, and the stage where they should optimally be applied (research, development, testing), need to be devised. Legitimate uses of the technology need to be weighed and a balance struck which restricts these as little as possible whilst effectively preventing the dangerous military uses. In parallel, *verification* methods have to be selected or designed.

If all four steps have been carried out, states should start negotiating and, ideally, agree on appropriate limits to prevent the respective technology from military/weapons use.

Independent Scientific Analysis of NLWs

To a significant extent, the first step of preventive arms control is decisive. Reliable knowledge about the properties and effects of weapons is a precondition for assessment of military-operational aspects as well as evaluation under various other criteria. For many reasons, such knowledge is not regularly and completely provided by either the planners or developers of new weapons technologies. Producers wish to protect commercial information and to advertise and sell their systems. Military institutions have a tradition of secrecy and compete for state money. Proponents may fear that the political decision process towards development and acquisition is hampered or delayed by full disclosure of all details and the ensuing intense debate on the pros and cons; some may not even be aware of the problems. All this leads to the danger that decisions are taken on the basis of promises and incomplete information. It should be mentioned that, among the NLWs-developing countries, the US, which has the largest programme, also provides by far the greatest transparency.

What is needed is independent scientific analysis, including technical and medical aspects. Such research should be open, should disclose all sources and publish all results, thereby allowing critical review, as is usual in public science. Therefore such study would be most effectively conducted in an academic setting and funded by institutions that have no vested interests in weapons development.

Such analysis may sometimes include original research. However, in many cases surveying existing literature will lead to valid results. Military-relevant information is often published as grey literature and therefore this will also have to be included. Scientific contacts within military establishments or contractors will also be useful. In some cases, simply digging up the complete history of earlier developments and tests can be helpful, as with the experience of the US Army and CIA tests with psychoactive drugs in the 1950s and 1960s[2] and also with similar activities in other countries. Where no data have been published, quantitative estimates will often be possible using basic knowledge about the field.

In the field of NLWs, there are several topics where such analysis is desperately needed in order to find what reality may lie behind the myths or to 'answer the speculation with real science'.[1] Table 5 lists several such areas and Alexander's book mentions more technologies for which hard facts would be desirable.[2] Well-founded analyses of the more exotic and speculative concepts (such as acoustic levitation of heavy stones) would also be useful.

If the international community finds that the application of force is justified in certain circumstances but that it should be kept at a minimum, then the weapons needed for such operations should be decided upon in a transparent process. The same holds true for legal force within societies. By providing the international public with hard information, independent

TABLE 5

NLWs: AREAS FOR INDEPENDENT SCIENTIFIC-TECHNICAL ANALYSIS

Low-kinetic impact	Superacidic/supercaustic chemicals
	Catalytic depolymerisation
Propagation bounded acoustic shocked beam	Viscosification agents
Multi-explosion blast-wave source	Liquid-metal embrittlement
Vortex-ring generator and propagation	Calmative agents
	Psychoactive drugs
Electro-shock weapons	
Explosive electro-magnetic pulse	Micro-organisms against humans, animals, plants
High-power microwave	Micro-organisms against material
	Biomimetic anti-material agents

This list is not exhaustive; psychological operations and information warfare are excluded.

scientists can do very useful work. However, funding institutions need to make a conscious choice towards these kinds of projects if they are to be allowed to devote their capabilities to such analyses.

Notes

1. Boyd D. The Search for Low Hanging Fruit: Recent Developments in Non-Lethal Technologies. In: Dando M, ed. *Non-Lethal Weapons: Technological and Operational Prospects*. Coulsdon: Jane's, 2000.
2. Alexander JB. *Future War*. New York: St. Martin's Press, 2000.
3. Lewer N, Schofield S. *Non-Lethal Weapons: A Fatal Attraction?* London: Zed Books, 1997.
4. Collins KR, Bowie DR. A History of Engine Defeat through Chemical Means. In: National Defense Industrial Association. *Non-Lethal Defense IV*. Tyson's Corner, VA: NDIA, 2000.
5. Barry J, Morganthau T. Soon, 'Phasers on Stun'. *Newsweek* 7 Feb. 1994: 26–8.
6. Kokoski R. Non-Lethal Weapons: A Case Study of New Technology Developments. In: *SIPRI Yearbook 1994*. Oxford: SIPRI/Oxford University Press, 1994: 367–86.
7. Aftergood S. The Soft-Kill Fallacy. *Bull Atom Sci* 1997; **50** (5): 40–45.
8. *Special Technologies for National Security*. Los Alamos, NM: Los Alamos National Laboratory, April 1993: 9.
9. Fulghum DA. US Weighs Use of Nonlethal Weapons in Serbia if UN Decides to Fight. *Aviation Week & Space Technology*, 17 Aug. 1992: 62.
10. US Global Strategy Council. *Non-Lethality: A Global Strategy Whitepaper*. Washington, DC: USGCS, 1992.
11. Toffler A, Toffler H. *War and Anti-War: Survival at the Dawn of the 21st Century*, Boston: Little, Brown, 1993.
12. Evancoe PR. Non-Lethal Technologies Enhance Warrior's Punch. *National Defense*, Dec. 1993: 26–9.
13. Bunker RJ, ed. *Nonlethal Weapons: Terms and References*. INSS Occasional Paper 15. Washington, DC: USAF Institute for National Security Studies, 1996.
14. Joseph B, Picat M, Barbier F. Liquid Metal Embrittlement: A State-Of-The-Art Appraisal. *Eur Phys J AP* 2000; **5**: 19–31.
15. Kiernan V. War over Weapons that Can't Kill. *New Sci* 11 Dec. 1993: 14–16.
16. Campbell JR. Defense against Biodegradation of Military Material. In: National Defense Industrial Association. *Non-Lethal Defense III*. Kayrek, ND: NDIA, 1998.
17. Coppernoll MA. The Nonlethal Weapons Debate. *Naval War College Review*, Spring 1999.
18. Pasternak D. Wonder Weapons – The Pentagon's Quest for Nonlethal Arms is Amazing. But is it Smart? *US News and World Report Magazine*, 7 July 1997.
19. Altmann J. *Acoustic Weapons – A Prospective Assessment: Propagation, and Effects of Strong Sound*. Ithaca, NY: Cornell University Peace Studies Program, 1999.
20. Altmann J. Acoustic Weapons: A Prospective Assessment. *Science and Global Security* 2001; **9**: 49–121.
21. Müller J, Protz R, Sepp G. *Nichtletale Waffen. Vol.II*. München: Daimler-Benz Aerospace, 1995.
22. Arkin W. Acoustic Anti-Personnel Weapons: An Inhumane Future? *Med Confl Surviv* 1997; **14**: 314–26.
23. Joint Non-Lethal Weapons Program. *1999 Annual Report*. Washington, DC: JNLWP, 1999: 19.
24. SARA Inc. MSDD (Multi-Sensory Distraction Device). In: National Defense Industrial Association. *Non-Lethal Defense IV*. Tyson's Corner, VA: NDIA, 2000.

25. Department of Defense. *Non-Lethal Weapons*. Directive No.3000 – 3. Washington, DC: DoD, 1996.
26. Bunker RJ. Non-Lethal Terms and Definitions. In: Dando M, ed. *Non-Lethal Weapons: Technological and Operational Prospects*. Coulsdon: Jane's, 2000; Appx.2.
27. Sautenet V. Legal Issues Concerning Military Use of Non-Lethal Weapons. *E Law – Murdoch University Electronic Journal of Law* 2000; 7: 2, http://www.murdoch. edu.au/elaw/issues/v7n2/sautenet72nf.html (8 Jan. 2001).
28. Lyell Lord. Non-Lethal Weapons. *Draft General Report*. North Atlantic Assembly, Science and Technology Committee, AP238, STC (97) 8, Sept. 1997.
29. Altmann J, Liebert W, *et al*. Preventive Arms Control as a Prerequisite for Conversion of Military-Related R & D. In: Reppy J, ed. *Conversion of Military R&D*. Basingstoke: Macmillan, 1998.
30. Personal communication with H.J. Knackmuss, Institut für Mikrobiologie, University of Stuttgart.

Perspectives and Implications for the Proliferation of Non-Lethal Weapons in the Context of Contemporary Conflict, Security Interests and Arms Control

NICK LEWER and TOBIAS FEAKIN

Characteristics of Contemporary Conflict

Whilst the security system of the cold war was characterized by bipolarity and ideological clarity, today the world has no clear-cut dividing lines or overriding threat.[1]

It is useful to start by briefly looking at some of the general trends of conflict and intervention patterns today, as this will help contextualize our later discussions on the revolution in military affairs, the requirements of modern armed forces and the place of non-lethal weapons technologies in meeting these challenges. In 1999 there were 27 major armed conflicts in 25 countries; of these only two were inter-state (Ethiopia-Eritrea and India-Pakistan over Kashmir), the rest being intra-state (including Algeria, Colombia, Chechnya, Indonesia-East Timor, Sri Lanka, Sudan and Kosovo). A major armed conflict is one which results in battle-related deaths of over 1,000 people in a year, where the conflict is about government and/or territory.[2] There is no reason to believe that this trend in intra-state conflicts will not continue for some time. Five intra-state conflicts attracted foreign military intervention:

Country	Intervenors
East Timor	United Nations led by Australia.
Democratic Republic of the Congo	Government side supported by troops from Angola, Namibia and Zimbabwe.
Republic of Congo	Congolese rebels supported by troops from Rwanda and Uganda.
Sierra Leone	ECOWAS, led by Nigeria.
Kosovo	NATO

Of these interventions it is interesting to note that two were 'sanctioned' by international organizations (Sierra Leone and East Timor), the NATO operation was 'tacitly' sanctioned but neither interventions in the DRC or Republic of Congo were internationally sanctioned. The states, both intervenors and conflict affected, lie along a spectrum that could be simply summarized as follows:

FIGURE 1
THE SPECTRUM OF STATES

STRONG STATES ⟺ CONTESTED STATES ⟺ WEAK STATES ⟺ FAILED STATES

⇕

AREAS OF STABILITY ⟺ AREAS OF INSTABILITY

The 'condition' of the state, the type of conflict and intervention response influences whether non-lethal weapons (NLWs) are seen to have military and/or political utility.

Non-State Groups

The dynamics of intra-state conflicts are made more complex by the presence of non-state armed groups such as liberation and terrorist organizations, criminal organizations (who often have a close relationship with corrupt officials in government), warlords and bandit gangs. Many of these have established international links. In these protracted social conflicts, leaders often manipulate emotive factors such as ethnicity and religion for their own personal gain. Such conflict entrepreneurs cloak their motives in a narrative of 'grievance' but their primary motives may be 'greed'[3] – less about politics than economics. Two types of violence may be broadly distinguished within the conflicts – 'top-down' and 'bottom-up'. Top-down violence refers to the mobilization and structural support given by leaders and entrepreneurs, whilst bottom-up violence is that perpetrated and embraced by ordinary people.[4,5] This apparent increasing use of violence to settle disputes is a key threat to global security and a clearer understanding of its nature and cause is needed – particularly when it presents in such barbaric and cruel ways. This needs to be placed within a broader historical context relating, for example, to the experience of nation/state building in Europe over the last few hundred years and the attempt at 'accelerated' nation building today. There are huge differences between nations in a post-modern condition of globalization and others that are in a condition of state building.

When high-tech armies, such as those of the United States or other western powers, intervene against non-state armed groups, the war is 'asymmetric'. Despite the sophisticated weaponry that such states possess, much combat is likely to be close, dirty and bloody.[6] As already mentioned, non-combatants (such as refugees) are mixed (sometimes deliberately) with combatants, often in densely populated urban areas.[7] One threat that is extremely difficult to counter is that of suicide terrorism, a tactic used by many groups all over the world. Especially deadly are the Liberation Tigers of Tamil Eelam (LTTE) in Sri Lanka and Hizbollah in Lebanon. According to Gunaratna:

> Terrorist groups are setting a dangerous trend of using suicide bombers to destroy targets far away from their theatres of war. Many groups are likely to use suicide bombers to infiltrate target countries and conduct suicide attacks against Western VIPs and critical infrastructure in the foreseeable future.[8]

Other security challenges come from sections of civil society who oppose global political and economic processes, such as those which fall within the concept of globalization. In the view of United Nations Secretary General, Kofi Annan:

> Globalisation offers great opportunities, but at present its benefits are very unevenly distributed while its costs are borne by all. This is the central challenge we face today, to ensure that globalisation becomes a positive force for all the world's people, instead of leaving billions of them behind in squalor.[9]

It is evident from the protests at recent World Trade Organization meetings that many around the world take the latter view of globalization as something which benefits the wealthy and advantaged and bypasses (and/or exploits) those most in need. These protestors generally employ non-violent methods but, as noted, a minority are prepared to use more violent approaches that still require non-lethal responses from police forces. Other internal threats to a state's national interests come from pressure and campaigning groups (a recent example is the disruption caused by a relatively small number of people over the cost of fuel in the United Kingdom), drug trafficking and disruption of computer networks (information/cyber warfare).

Whilst we have highlighted a few of the characteristics of contemporary conflict, it should not be forgotten that a state's armed forces still have to be prepared for 'traditional' inter-state war.

Non-Lethal Developments

Non-lethal weapons will be able to make a major contribution to the capability of US military forces across the entire spectrum of conflict including protection of observers, prevention of terrorism at home and abroad, peacekeeping and peace enforcement, and large scale military operations.[10]

NLWs are explicitly designed and primarily employed to incapacitate personnel or material whilst minimizing collateral damage to property and the environment.[11] The first-generation weapons have been extensively described elsewhere and include rubber and plastic baton rounds, chemical sprays, electrical stunning devices, vehicle barriers and electronic and computer applications. First-generation NLWs are readily available in many parts of the world. The development of second-generation weapons will proceed because of a variety of influences. These include response to western public demand for fewer casualties (both in military and civil operations); a call for less collateral damage, a high regard for life in the west, environmental concerns over weapons 'pollution', the changing nature of warfare (for example, an increase in military operations in urban terrain and asymmetric warfare), peacekeeping (peace support) operations, the push from advances in non-lethal technologies and political and commercial lobbying. All these factors are drivers for proliferation. Second-generation weapons will have certain characteristics which are qualitative improvements on their predecessors: they will be dual capable; deliverable from a greater distance with more accuracy (from unmanned aerial vehicles [UAV] platforms or cruise missiles); more subtle in their effect (especially those NLWs associated with psychochemistry and neuroscience); smaller, more energy efficient and easier to transport and more environmentally friendly.

Proliferation Trends

As second and third generation weapons are developed, current NLWs will proliferate. Further difficulty is that non-proliferation measures will be difficult to implement since the technologies and equipment are not unique to non-lethal technologies.[12]

A few general introductory remarks may be helpful to lead us into a more focused discussion about the proliferation and control of NLWs. Should the proliferation of NLWs be stopped or controlled and who wants to do this? Since the end of the Second World War considerable effort and resources have been put into arms control and anti-proliferation measures (conventional weapons and weapons of mass destruction) with varying degrees of success. Whilst progress has been made, there are still serious problems with the Nuclear Non-Proliferation Treaty (where, for example, two non-signatories, India and Pakistan, have both developed nuclear weapons), the Anti-Ballistic Missile Treaty, the Comprehensive Test Ban Treaty and START II talks. The perennial problem for any treaty is to give it effective enforcement teeth. Signatories to a treaty who renege can always be 'named and shamed'. There are also UN arms embargoes on weapons to certain 'rogue' states which may be mandatory (Iraq, Somalia and Liberia) or non-mandatory (Afghanistan, Ethiopia and Eritrea). Sceptics may question whether arms control really prevents those who are determined to develop or acquire them and others argue that it is legitimate to develop counter-measures to actual and possible threats to national security – therefore there is a need to know about the weapons themselves.

For our purposes the concepts of proliferation and the arms trade are linked. Proliferation has generally been used to mean the transfer of technology, equipment, knowledge and strategic goods to countries who do not possess them.[13] Two broad patterns of proliferation have been used to describe the spread of weapons – horizontal and vertical. 'Horizontal proliferation' was taken to describe the spread of weapons to other countries or areas, whilst 'vertical' was used when talking about quantitative or qualitative advances of weapons within a state. More recent terminology speaks of 'armament dynamic', 'supply side' and 'demand side' aspects of proliferation.

- *Armament Dynamic* – This is the total process of developing, acquiring and maintaining a particular type of weaponry for the armed forces as well as developing the necessary procedures to integrate that weaponry into military doctrine.

- *Supply-side Proliferation* – This is the flow of technology, equipment and knowledge from states possessing these commodities to states lacking them and is determined by geopolitical, bureaucratic, economic and technological motivations. In some cases the supplier state will try to control transfers through export control policies, diplomacy and other political means.

Of particular interest here are the first-generation NLWs, since it is these older technologies which are most likely to be exported or manufactured under licence, whilst the newer non-lethal technologies will be more closely guarded for reasons of national security, technological advantage over opponents and commercial protection.

- *Demand-side Proliferation* – This occurs when a state decides to obtain a weapon when such a capability does not yet exist, provided this decision is followed by an armament dynamic. Some demand-side proliferation is covert in order to keep one step ahead of opponents. In the case of NLWs, demand-side proliferation is determined by both domestic and strategic issues. At the domestic level there are needs for weapons to manage and cope with riots, acts of terrorism, crime and anti-guerrilla warfare. Non-proliferation processes may impede domestic law-enforcement policy and operations. Strategically, NLWs may be developed as part of force projection in war-fighting and peace support interventions. In any case states may want to develop their industrial and technological capacity so that they are not reliant on outside sources for particular weapons systems.

- *'Official' Arms Trade and the Black Market* – The export side of weapons sales forms an important part of the economic and employment context of many nations. Selling weapons also enables the sellers to a have economic and political 'stakes' in the recipient country.

Why Should We Want to Control the Proliferation of Non-Lethal Weapons?

We need to ask why states/organization want to acquire a non-lethal capacity. The motivations are many and relate to the level of intervention.[14] The previous section has described a framework within which to discuss proliferation issues. NLWs proliferation is also linked to other weapons developments (spin-offs), research into counter measures, the state of the military-industrial complex, research laboratory pressure, domestic lobbies, the revolution in military affairs and the military technology revolution.

Concerns about NLWs proliferation centre around key issues relating to biomedical concern, misuse (torture and punishment), use for political control and suppression, damage to the environment, blurring of civil and military operations, infringement of personal privacy and implications for conventions and treaties. The Omega Report[15] found that 110 countries world-wide deploy non-lethal riot control weapons and that in 47 of these countries they are used not only for crowd and riot control 'but also in conjunction with lethal force rather than as a substitute for it, leading directly to injury and fatalities'. It noted that there was very little biomedical research being done on the effectiveness of NLWs or their biological effect on targets. Regarding chemical NLWs, the Omega Report suggested that research into the safety of crowd control irritants should be placed on public record before authorization of deployment is given. In the US, the Human Effects Advisory Panel (HEAP) at Pennsylvania State University was commissioned by the Joint Non-Lethal Weapons Directorate (JNLWD) to provide an independent assessment of human effects issues surrounding non-lethal weapons. HEAP was critical of the services' methodology for assessing injury from blunt impact munitions because the models used were not validated and did not address a number of likely injury types. The report also noted that none of the models used for blunt trauma looked at the concept of 'minimal non-lethal effect'.[11] The JNLWD state that:

> fielded weapon systems must be accompanied by a comprehensive package of data regarding health effects to users and bystanders, health hazards to targeted individuals, and weapon effectiveness.[11]

The use of non-lethal weapons for torture and punishment has also been well-documented.[16-18] Omega identified 33 countries in which crowd control weapons had been used to facilitate human rights violations.[19] International codes of conduct and mechanisms to prevent non-lethal technologies from being used to violate human rights are desperately needed. EU countries frequently export NLWs to countries that are known to use them for torture. The European Union Code of Conduct on Arms Exports (1998) states that export licences will not be given if weapons to be used for internal repression or if they provoke or prolong armed conflict. There have also been examples of non-lethal technologies, designed to restrain or prevent violence and/or escape, being used as punishment:

> a criminal who kept talking in court was administered an electric shock (50,000 volts) by a Judge. He was wearing a stun-belt because he had a previous record of violence and disrupting court procedures. One company reported that their model of the stun-belt had been used 27 times – eight of them accidental.[17]

A watching brief is required on the role of 'privatized' security and prison organizations employing NLWs to control prisoners and/or quell disturbances. There are obvious implications with respect to dangers of misuse and unaccountability in what is becoming an increasingly privatized prison (and, in some parts of the world, police) system.

The structure of an intervening force may be a combination of combat troops and civil police units with different mandates and operational requirements. There is a concern about the militarization of civil policing techniques and methods through the development of dual-use technology and training and a blurring of civil and military applications. US Attorney-General, Janet Reno, acknowledged this in 1994 when the Department of Defense and the Department of Justice planned to develop dual-use NLWs through a Centre for Defense and Law Enforcement Technology:

> When police use these devices they must be constrained by the knowledge that the people they are restraining aren't enemies; they are fellow citizens, with a full set of civil rights.[20]

Whilst there is a general acceptance that there needs to be greater complementarity and co-ordination between military and civilian agencies in policing and humanitarian crises, there still remains a wide gulf between the mindsets, rationale and motivations of the two types of organization. These differences influence how and when the use of NLWs may be recommended and who should use them.

Tactical NLWs are now being more aggressively marketed – as a quick search on the Internet illustrates – and private companies now offer a wide-range of 'personal protection' devices. Such weapons have an obvious application in criminal activity. There is a tension between giving people access to what they may perceive as legitimate means of self-protection and the acquisition of NLWs by criminals for assault and robbery. Some would use an argument similar to that for arms and gun control in the US:

> In short, the nasty, lying, thieving underworld corrupts an otherwise perfectly legitimate, licit, even beneficial (for US Security) arms trade. Stop the bad guys and the problem goes away.[21]

Because NLWs span a range of weapons technologies they impact on several treaties, particularly the Chemical and Biological Weapons Conventions and the UN Inhumane Weapons Convention. In the face of new non-lethal technologies and a changing global security system, must we re-conceptualize the very nature of arms control and proliferation with respect to NLWs? Non-government organizations and campaign groups hold an important role in promoting treaties and conventions and recent occasions where public support has been mobilized to control weapons include legislation concerning blinding laser weapons and landmines.

Reasons for More NLWs

The previous sections have highlighted some of the arguments put forward for the non-proliferation of NLWs but there are strong arguments for wider development and proliferation of non-lethal technologies. Whilst pointing out the cautions associated with the use of non-lethality, Lovelace and Metz also support their utility:

> Non-lethality could provide political decision-makers and military commanders with means to dominate the portion of the spectrum of force that lies between diplomacy and lethality. In doing so, they will be better able to apply the precise psychological pressure required to modify an adversary's behaviour in a certain way. Non-lethality can be used to deter or pre-empt conflict, separate belligerents and allow for 'cooling off', encourage negotiation, protect non-combatants, facilitate disaster relief and humanitarian assistance operations, enhance the effectiveness of lethal weapons and other instruments of national power, and reduce risks to US forces.[22]

There are obvious advantages to arming civil police with a range of NLWs. For example, one could argue that it is better to be momentarily stunned by a Taser gun than bashed about the head with a truncheon and that it is better to be disarmed by CS gas than by a bullet from a revolver. While arming police non-lethally helps build an ethos within which the police operate, less violent techniques for crime prevention and management have also to be developed.

Case Study – India

Tear gas was first used in India by the British to help control large-scale rioting. Until the late 1970s India imported all of its supplies of chemical riot control munitions from the US, UK and France. However, research was carried out by the Bureau for Police Research and Development into the development and manufacture of chemical control agents[23] and India now has its own tear gas munitions factory.

The Tear Smoke Unit

The Border Security Force (BSF), set up in 1965, is the largest paramilitary force in India. It supports army operations in areas such as the Kashmiri border. In 1976 the BSF set up the Tear Smoke Unit. This was driven by both operational and commercial factors, since the cost of having to import crowd control munitions (mostly from the US) was becoming prohibitive.[24] It was also felt that an internal manufacturing capacity was strategically important. The BSF factory is situated at Tekanpur, Gwalior, in the Madhya Pradesh region and is one of only two tear gas factories in South Asia (the other being in South Korea). This factory is unique in that the entire factory workforce is recruited from the ranks of the BSF. Original chemical products were manufactured using CN gas contained in either a basic smoke grenade form or as a gun cartridge. However, the range of non-lethal weapons has grown considerably, utilizing CN, CR and CS gases, and includes:

- *Tear Smoke Shells* – These come in either a metal or plastic coating. The body, which is made of plastic, melts when the shell bursts so that it is almost impossible for targets to pick it up and throw it back. The shell emits tear gas for 20–40 seconds after a delay of four to six seconds.

- MK IV Tear Smoke Grenade *'Antiriot'* – Also made with the melting plastic casing, the grenade emits gas for 30–50 seconds after a delay of a second.

- 3-Way Tear Smoke Grenade *'Antiriot'* – This grenade is made of three segments which separate when the grenade bursts. Each segment emits gas for 20–40 seconds after a delay of two seconds.

- Dye Marker Grenade *'Amit'* – This is a basic dye grenade which sprays an indelible coloured dye over rioters. The dye remains on a person for up to 48 hours.

- Two-in-One Shell *'Pravir'* – This shell contains two segments; one is filled with gas and the other has a 'flash bang' effect.

- Stinger Grenade *'Vrishchak'* – On explosion these grenades combine a 'flash bang' effect with the dispersion of hundreds of small hardened plastic pellets.

- Wood Piercing Shell *'Rudra'* – This shell was developed for siege situations. Fired from a gas gun, it can penetrate wood up to an inch thick from a range of 60–70 metres.

- Self Protection Aerosol *'Rakshak'* – This is a small handheld device filled with oil of capsicum. It comes in various sizes and has been designed to look like a perfume bottle.

- *Electric Tear Smoke Grenade* – This was developed to aid perimeter defence. It can be triggered either by a trip-wire or by remote control.

- Stun Grenade MK III *'Naagpash'* – Another perimeter-defence weapon containing a 'flash bang' charge. It is electrically initiated, with a battery-operated triggering mechanism.

- Floating Tear Gas Device '*Magar*' – This device is designed to float in water and can be used, for example, to incapacitating criminals in boats or for the protection of offshore installations. Once triggered, it discharges large volumes of gas for 50–60 seconds.

- Other weapons in development at the factory include the '*Super Tear Gas*' grenade, a three-way grenade which separates in the air after being thrown. Each segment contains a different type of tear gas, one CN, one CS and one CR.

The factory also produces various plastic and rubber bullets and has a sister company, owned by the BSF, which produces the firing mechanisms for all of these weapons. The factory at Gwalior now supplies all of India's police and paramilitary forces with tear gas munitions and is beginning to supply the army who are showing a particular interest in mortar-fired projectiles. However, the Indian army, which has historically only been armed with lethal weapons, is starting to look at non-lethal options. Researchers at the factory are exploring the development of more blast dispersion shells and mortar fired projectiles.

India's Deployment of Non-Lethal Weapons

The 'nuclear threat' from Pakistan is central to India's security concerns and the development of a nuclear force has demanded huge military resources. India also maintains a large standing army, capable of fighting a sustained conventional war. Within this current security context, developing a role for NLWs within the Indian Army is of low priority. Weaponry procurement is directly related to national policy and non-lethal weapons do not fit in with India's present national policy and doctrine. This is not to say that the Indian Army never deploys non-lethal weapons – its National Security Guard (NSG) use various non-lethal weapons when dealing with, for example, terrorist incidents. The NSG has the usual non-lethal armoury available, including tear gas munitions and flash bang grenades.

The greatest application of NLWs is in India's paramilitary forces. These forces are intended to back both the army and local police forces when necessary. It is this 'multi-tasking' that has led the Indian authorities to investigate non-lethal options which can be used in both military and policing. The newest of India's paramilitary groups is the Rapid Action Force (RAF), founded in 1992, which is a specialist force designed to deal with serious crowd control problems and the protection of VIPs. It was formed in response to the increase in communal riots and civil unrest in India during the 1980s. Officially part of the Central Reserve Police Force (CRPF), it draws all of its recruits from the ranks of the CRPF but functions independently in terms of the tasks it carries out. What is unique about the RAF is that its forces are heavily armed with NLWs and their rules of engagement state that live ammunition can only be used as an absolute last resort. The non-lethal arsenal includes rubber and plastic bullets, various types of CS and CN gas canisters, electro-shock batons and stun guns, rubber side-handled truncheons, plastic lathes (a 3–4ft stick usually made from cane), plastic shields, body armour and protective helmets. It was this almost exclusive reliance on a non-lethal armoury that led the United Nations to request the RAF's services in helping to police Kosovo in summer 2000.

The Proliferation Cycle in India

As seen above, India has developed its own NLW manufacturing infrastructure (demand-side proliferation followed by an armament dynamic). By 2000 India was exporting a wide-range of its tear gas munitions in South Asia (supply-side proliferation). Details are classified by the Indian Government, but it is known that a range of NLWs were sold to Mauritius in March 2000. The BSF factory would like to increase its export base, but the Indian Government is currently constraining this. Since the products produced at the BSF factory are only a third of the price of their western equivalents, there is real potential for large-scale exportation that would lead to considerable expansion of production at the factory and both demand-and supply-side proliferation.

Conclusions

The rationale and arguments for the proliferation or non-proliferation of NLWs are complex and multi-faceted. This chapter has only laid out a preliminary framework within which the issues may be considered in more detail. NLWs are a small sector in the armaments industry but they will play an increasingly important part in security considerations within the contexts of technological advances, police and military operational requirements, political demands and humanitarian influences. Proliferation is inevitable. The development of NLWs in India illustrates this proliferation cycle – supply, demand, armaments dynamic and assimilation. Because of the very nature of NLWs, important ethical issues and debates cannot be avoided. Despite their apparent benign nature, NLWs can have a profound malign effect in areas such as torture, the infringement of personal freedom and impact on the environment.

More detailed control and transparency proliferation mechanisms are required, particularly with regard to states which are known for their human rights abuses. Only open and extensive debate can draw the line between legitimate application of NLWs and their misuse.

Notes

1. Rotfield A. In Search of a Global Security System for the 21st Century. In: Stockholm International Peace Research Institute (SIPRI). *SIPRI Yearbook 2000: Armaments, Disarmament and International Security*. Oxford: Oxford University Press, 2000: 1–13.
2. Stockholm International Peace Research Institute (SIPRI). *SIPRI Yearbook 2000: Armaments, Disarmament and International Security*. Oxford: Oxford University Press, 2000: 15.
3. Collier P. Doing Well Out of War: An Economic Perspective. In: Berdal M, Malone D. *Greed and Grievance: Economic Agendas in Civil Wars*. London: Lynne Reinner, 2000.
4. Goodhand J, Hulme D, *et al*. Social Capital and the Political Economy of Violence: A Case Study of Sri Lanka, *Disasters* 2000; **24**: 390–407.
5. Keen D. Incentives and Disincentives for Violence. In: Berdal M, Malone D. *Greed and Grievance: Economic Agendas in Civil Wars*. London: Lynne Reinner, 2000.
6. Metz S. *Armed Conflict in the 21st Century: The Information Revolution and Post-Modern Warfare*. Carlisle, PA: US Army War College, 2000.
7. Bowden M. *Black Hawk Down*. London: Bantam Press, 1999.
8. Gunaratna R. Suicide Terrorism: A Global Threat. *Jane's Intelligence Review*, 20 Oct. 2000, http://www.janes.com/security/regional_security/news/usscole/jir001020_1_n.shtml.
9. Annan K. *We, The Peoples: The Role of the United Nations in the 21st Century*. Millennium Report of the Secretary-General of the United Nations. New York: 2000 (Sections 13–14 and 18–65), http://www.un.org/millenium/sg/report/full.htm.
10. Council on Foreign Relations. *Nonlethal Technologies: Progress and Prospects*. Washington, DC: Independent Task Force on Non-Lethal Weapons, Council on Foreign Relations, 1999.
11. Joint Non-Lethal Weapons Directorate (JNLWD). *Newsletter* 1999, **2** (3): 1.
12. Siniscalchi J. *Non-Lethal Technologies: Implications for Military Strategy*. Occasional Paper No.3. Alabama: Center for Strategy and Technology, Air War College, Air University, Maxwell Air Force Base, 1998.
13. SIPRI. At: http://www.cbw.spiri.se/cbw/.
14. Lewer N. Benign Intervention and Non-Lethality: Wishful Thinking for the 21st Century. In: Dando M, ed. *Non-Lethal Weapons: Technological and Operational Prospects*. London: Jane's, 2000.
15. Omega Foundation. *Crowd Control Technologies: An Assessment of Crowd Control Technology Options for the European Union*. Luxembourg: European Parliament, Directorate General for Research, Directorate A, Division of Industry, Research and Energy, Scientific and Technological Options Assessment (STOA), 2000.
16. Amnesty International. *Arming the Torturers: Electro-Shock Torture and the Spread of Stun Technology*. London: Amnesty International, March 1997.
17. Kettle M. 50,000 volts for talking in court. *Amnesty International Medical Group Newsletter* 1998; **10** (3): 2.
18. Omega Foundation. *An Appraisal of the Technologies of Political Control*. Luxembourg: European Parliament, Directorate General for Research, Directorate A, Division of Industry, Research and Energy, Scientific and Technological Options Assessment (STOA), 1998.

19. Ibid.: 52.
20. Pemberton M. Robocop IV: Dual-Use Comes to Police Work. *New Economy*, Summer 1994: 11.
21. Nelson D. Damage Control. *Bull Atom Sci* 1999; 55 (1): 55–7.
22. Lovelace D, Metz S. *Nonlethality and American Land Power: Strategic Context and Operational Concepts*. Carlisle, PA: US Army War College, 1998.
23. Swearengen T. *Tear Gas Munitions*. Springfield, IL, 1996.
24. Ghosh S, Rustamji K. *Encyclopaedia of Police in India*. Vol.1. New Delhi: Ashish, 1993: 63.

Non-Lethal Weapons:
R²IPE for Arms Control Measures?

VICTOR WALLACE

> Our Commanders need and want a non-lethal capability that they
> can easily and quickly tune up or down, or switch quickly to a lethal
> effect.

Lieutenant General Martin R. Steele (USMC)[1]

The rapid growth of western military involvement in humanitarian
interventions and peace support operations (PSO) over recent years has
prompted significant political and military interest in 'new tools for new
jobs'.[2] Attention has focused specifically on the potential utility of novel
equipment and weapon systems in complementing intervention actions in
semi- or non-permissive conflict environments. Moreover, military
intervention concepts and doctrine, particularly related to low intensity
conflict, have undergone radical change.[3] 'Non-lethal' weapon (NLW)
systems have been viewed with particular attraction in some political and
military circles because of their potential to contribute to the attainment of
tactical, operational and strategic objectives, but at the same time
minimizing casualties and collateral damage and ensuring the protection,
cohesion and legitimacy of humanitarian/PSO forces.[4] This potential is seen
to be important for two principal reasons.

In the first instance, the use of 'lethal' force carries the risk of
undermining the status and impartiality of intervening military forces. This
in turn can weaken the influence of, and popular support for, intervening
forces among belligerent parties and, consequently, undermine their
'consent'.

Second, maintaining the legitimacy of intervening forces has become a
major consideration for western governments[5] and military planners[6]
because of the need to sustain both domestic and international support for
such operations once troops have been committed. Consequently, when
intervening to protect less-than-vital, humanitarian interests, a heavy
premium is placed on applying military force in a proportionate manner in
order to minimize casualties and collateral damage.

Conceptually, NLW advocates have proposed a 'force continuum' that balances the threat and use of lethal force, underpinned by flexible rules of engagement (ROE). It has been argued that NLWs complement this concept and provide three distinct advantages: they are humane; they provide commanders with greater control of deteriorating situations; and they are less likely to provoke retaliatory action or cause public outcry.[7] The theory suggests that NLW technology could potentially provide military commanders and politicians with a 'silver bullet' when intervening in semi- or non-permissive environments.

However, medical evidence would suggest otherwise. In 1991, Garfield and Neuget[8] published a systematic review and analysis of the effects of armed conflict on both military and civilian populations over a period of 200 years. The evidence suggested that mortality could be linked directly to changes in military technology and that morbidity has been both physical and, increasingly, psychological involving a greater proportion of the civilian population.[9] Moreover, overall mortality rates associated with armed conflict have remained relatively constant since the Second World War,[10] notwithstanding the reduction in force numbers and evolving concepts of lethality, concentration and mass.[11]

In low intensity conflict, the evidence suggests that the number of people killed may be greater than the number wounded when firearms are used against individuals who are immobilized, incapable of defending themselves or find themselves in a confined space.[12] This point is reinforced by evidence of mortality associated with the use of non-lethal kinetic weapons such as batons and rubber bullets in civil law enforcement[13] and the fact that the availability of such weapons does not affect the rate of killings by police forces.[14]

Yet, despite this historical evidence, the drive towards greater technical innovation to address both precision and discrimination continues. The industrial sector, particularly in the United States, has led this technological revolution. One example is the concept of the 'smart bullet' where in future guided rounds could 'allow for improved lethality by aim point selection' and enable the user 'to choose between an arbitrary number of types, from armor penetrating to "non-lethal"'.[15]

Despite such claims of efficacy, a weapon system must be analyzed against its effects. Logically, effects and outcome will determine not only a weapon's utility, but also its legality. But how does one quantify effect or outcome? Blinding laser systems and anti-personnel mines have both been assessed and banned on the basis of irreversible retinal damage and indiscriminate injury, respectively.[16] In addressing the issue of legality, recent medical literature has raised the principle of 'superfluous injury and unnecessary suffering'[17] – outlined in both the Geneva Convention[18] and Hague Regulations[19] and endorsed in 1996 by the International

Court of Justice. Moreover, definition of this principle has been proposed as a measure of a weapon's effect and, therefore, a potential determinant of the legality of novel weapon systems.[20]

This chapter assesses the utility, effects and legality of NLW systems and questions whether they should be subject to specific arms control measures. The issue of utility is addressed by examining NLW policy and doctrinal developments in the context of the revolution in military affairs (RMA), lessons identified by US Marine Corps (USMC) involvement in Somalia and NLW evaluation methods. Moreover, a wide range of NLW technologies are examined in light of the principle of 'superfluous injury or unnecessary suffering'. The chapter concludes by suggesting the R²IPE acronym (Reversibility of effect – 'Rheostat' capability – Information – Policy – Effect) as an additional assessment tool for considering whether to develop and acquire a particular NLW system.

Non-Lethal Weapons Utility

The first difficulty when considering the utility of NLW systems involves terminology.

Terminology and the RMA

The military and civil law-enforcement communities have wide experience in the employment of baton rounds, water cannon and riot control agents such as CS gas. However, it is important to note that NLW *per se* are not new. In the 1940s, German scientists experimented with 'vortex' technology comprising high-energy acoustic weapons designed to disable or destroy men and material.[21] Moreover, between 1961 and 1965 American forces used the defoliant 'Agent Orange' over dense jungle canopies as an area denial weapon against North Vietnamese troops.[22] However, the long-term effects of the agent included both cancer and congenital malformations.[23] Neither of these weapon systems was described as 'non-lethal' in their historical context, yet both highlight the inherent difficulties faced by contemporary political and military planners in attempting to define exactly what constitutes 'non-lethal' and to develop policy and doctrine to guide the use of NLW.

At first sight the term 'non-lethal' appears to suggest the potential to inflict zero fatalities. However, this proposition sits uncomfortably with the purpose of a weapon, which can be defined as an instrument specifically designed to cause bodily harm. Equally, NLW systems can be utilized against equipment and infrastructure, not just individuals. A range of NLW technologies and their potential utility in future symmetric or asymmetric conflicts is summarized in Figure 1.

FIGURE 1
UTILITY OF SOME NLW TECHNOLOGIES IN FUTURE CONFLICTS

Level of Command ⇩	Future Conflict ⇒	War between developed nations using high technology weapons	War between a developed nation and irregular forces
STRATEGIC (the use of force to achieve political objectives)		Voice synthesis Computer viruses Carbon ribbon	Deception Fuel additives Voice synthesis
OPERATIONAL (the use of force to degrade or defeat military forces)		High-energy lasers Microwaves Combustion modifiers	Lasers Soil destabilisation Non-nuclear electro-magnetic pulse weapons
TACTICAL (the use of force to defeat or destroy the enemy's warfighting capability)		Lasers Obscurants Optical coatings Anti-traction agents	Noise/odour/lights Nets/entanglers Infra and ultra sound Tyre attack

Taken from Lt.-Col. A. Roland-Price, 'Can Future War Be Non-Lethal', *British Army Review* 113, Aug. 1996: 23.

NLW technologies can have both direct and indirect effects of varying severity on a target population. These effects can be brought to bear against the individual or collectively, the latter through the disruption of utilities to key facilities such as hospitals or organizations such as the emergency services. Moreover, anti-material NLW such as obscurants or combustion modifiers dispersed by aircraft or unmanned aerial vehicles may not be discriminate in their effects on a target population or the environment.

To address the problem of definition, additional terms have been proposed. Definitions such as 'pre-lethal', 'less-than-lethal' and 'worse-than-lethal' have been developed in an effort to quantify a particular weapon's immediate and delayed effects over time.[24] Crucially, however, proponents of NLW have emphasized the 'reversibility' or minimal nature of their effects, thus attempting to make a distinction between NLW and conventional weapons.[25] The following policy statement on non-lethal weapons released by NATO on 13 October 1999 enshrines this emphasis:

> Non-lethal weapons are weapons which are explicitly designed and developed to incapacitate or repel personnel, with a low probability of fatality or permanent injury, or to disable equipment, with minimal undesired damage or impact on the environment.[26]

The same policy statement also identified the following areas where NLWs should enhance the capability of NATO military forces:

- To accomplish military missions and tasks in situations and conditions where the use of lethal force, although not prohibited, may not be necessary or desired;
- To discourage, delay, prevent or respond to hostile activities;
- To limit or control escalation;
- To improve force protection;
- To repel or temporarily incapacitate personnel;
- To disable equipment or facilities;
- To help to decrease the post-conflict costs of reconstruction.

The NATO statement is pertinent for two reasons in the context of this study. First, there is an acceptance that NLWs are not 'non-lethal' and do carry a low probability of fatality or injury. Second, the intent is to minimize, rather than to negate, the disability to individuals, equipment and the environment. However, while articulating the potential utility of NLWs, the policy does not 'quantify' the nature or duration of NLWs' effects. Notwithstanding these issues, NLWs have been advanced as evidence of a military technological revolution and provide the technical means to develop new concepts of warfare.[27]

Discussing the broader utility of NLWs, Lieutenant General Anthony Zinni (USMC)[28] has argued that NLW can provide a capability that fulfils the Sun Tzu ideal.[29] According to Sun Tzu, 'generally in war the best policy is to take a state intact; to ruin it is inferior to this'.[30] This is a laudable view. However, the interpretation of Sun Tzu is open to question. The NLW system in itself may not be the panacea. Sun Tzu emphasized the key principles and priorities of offensive action as: attacking the enemy's strategy, including early management of crises before they occur; attacking the enemy's plans; disrupting alliances; and, finally, attacking his army.[31] The question remains: can the utility of NLWs contribute to a potentially beneficial evolution in warfare within the RMA?

Several authors have examined the concept of 'benign intervention' particularly in a peace enforcement environment.[32] Again, emphasis is placed on the 'reversibility' of effect against individuals, property and infrastructure in order to maintain civil confidence by limiting casualties and collateral damage and, thus, minimizing the financial and manpower costs of rebuilding the target state. Yet, current military doctrine is not structured to accommodate mutually exclusive 'lethal' and 'non-lethal' force. A true RMA would require radical changes to occur in this area. However, in 'complex emergencies' of the future NLWs may provide a 'quick fix' in the face of aggression but in the process undermine other recognized PSO principles such as communication, confidence building and negotiation. In short, a military force relying on NLWs alone may subdue the immediate crisis, but ultimately distance itself from the underlying problems.

This criticism was highlighted when the USMC deployed NLW in 1995 during the United Nations (UN) withdrawal from Somalia in Operation 'United Shield'. According to Stanton, for example, 'Zinni going in person to talk turkey with Aideed had more to do with the relatively peaceful execution of United Shield than any non-lethal technologies ever did'.[33]

Moreover, the deployment and use of NLWs will have varying degrees of significance for the target population/leadership, which will depend upon cultural, social and economic factors. For example, the deployment of NLWs could be interpreted by a target population/leadership as a lack of resolve that could ultimately make the use of lethal force against and by peacekeepers more, rather than less, likely. As Ralph Peters argued in 1994, 'We will face opponents for whom treachery is routine, and they will not be impressed by tepid shows of force with restrictive rules of engagement'.[34] The situation may also be exacerbated by the media which forms an important and rapid conduit for information to both friendly and opposition forces.[35]

Furthermore, there is the potential for peacekeepers to be drawn into manufactured situations where they are forced to use lethal force.[36] Subsequent claims by a target leadership that an incident was 'a peaceful protest' could be used to exploit a tactical situation for operational and strategic effect. Allegations that peacekeeping forces should have used NLWs may also be levelled, thereby calling into question the proportionality of their response.

Nevertheless, exponents of NLWs in the United States believe they provide a necessary and credible capability in the post-cold war era.[37] In particular, it has been proposed that NLWs fill the gap between the threat and application of lethal force in low intensity conflict conducted in urban environments.[38] However, 'non-lethal' and 'lethal' force are not mutually exclusive and need to be viewed as complementary. To highlight these issues further, the chapter proceeds by examining lessons derived from the US experience of deploying NLWs in Somalia.

Somalia, 1993–95

In May 1993, the UN initiated Operation 'Restore Hope' in an effort to supply humanitarian aid to Somalia, a country in the throes of civil war amidst starvation and a breakdown in civil law and order. Several countries – including the United States, Canada and Italy – offered troops to the humanitarian relief effort. Virtually all of the risks and difficulties associated with operating in a complex emergency were evident in Somalia. However, the United Nations subsequently undertook an ambitious programme attempting to rebuild the shattered nation and to undertake a phased programme of disarmament. Undermanned and under-equipped for the task, the UN force found itself in a complex crisis characterized by urban guerrilla warfare. By the end of 1994, more than 130 UN troops had died attempting to carry out the United Nations' mandate at a financial cost of some $2 billion.[39]

In the aftermath of renewed fighting in the Somali capital Mogadishu and the perceived failure of the mission in late 1994, the UN requested American assistance to support the withdrawal of UN forces from the area. US Central Command (CENTCOM) subsequently directed the 1st US Marine Expeditionary Force (MEF) – which had formed the command element of Operation 'Restore Hope' two years earlier – to take charge of the withdrawal operation, titled 'United Shield'. In order to minimize casualties and to reduce confrontation, Lieutenant General Zinni tasked his staff with pursuing 'less lethal alternatives' to deal with unarmed hostile elements in Mogadishu.[40] Zinni felt NLWs provided a riot control capability that minimized the risks to his own troops but, more importantly, plugged a gap in ROE; an 'operational niche in which our adversaries could be neither deterred or controlled ... against which we were unwilling to apply lethal force'.[41]

However, Zinni's staff encountered problems with NLW availability, quantity, field performance and training because of the novelty of the weapons systems. In addition, there were no specialized delivery means; the standard organic USMC inventory – such as the M203 rifle and grenade launching system – would have to be used. Moreover, discrepancies existed between the US government and the military in terms of NLW policy, doctrine and training.

Of particular concern to the US government was the fact that NLWs such as sticky foam could prove lethal. Furthermore, the possibility existed that American personnel could themselves become entangled. The US Department of Defense (DOD) approval process was delayed because of concern over the potential impact of media images showing individuals in a crowd or US servicemen suffocating under foam. However, the USMC never planned to use foams as an anti-personnel option; they were only to be used as barriers or area denial weapons.[42]

Flexible ROE were developed for the mission that incorporated NLWs and adhered to the principle of 'graduated response', but which incorporated the right to use lethal force if required.[43] Conceptually, this was described as a 'force continuum' allowing for use of the minimal force required in responding to a threat. NLWs such as sticky foam were included at the low end of the continuum, which then evolved to incorporate 40mm wooden batons, sponge grenades and beanbags towards the other end. As Lorenz pointed out, ROE incorporating NLWs were developed specifically for the mission. Since the introduction of NLWs presented problems in both interpretation and context, the issue of the ROE was delayed for more than two weeks by the Joint Chiefs of Staff approval process.[43]

The media played an active role during the deployment. Prior to the withdrawal operation, the first MEF training programme conducted offshore attracted huge media interest. However, Zinni and his team were concerned about the possibility that rival factions in Mogadishu would also watch the broadcasts and develop crude countermeasures. Furthermore, critics of NLWs highlighted the potential difficulties raised by crowd dynamics and the tactical environment in Somalia.[44] For example, a contrived, carefully organized riot involving large numbers of people with armed opponents mingling with unarmed women and children – conducted in a dense urban environment – could have made proportionate and discriminate response almost impossible. However, the withdrawal of UN forces subsequently took place almost unhindered with no loss of life. Yet, NLWs were not fully utilized except for the use of sticky foam for perimeter protection. Nevertheless, supporters of NLWs claim that the mission proved successful and that NLWs provided a credible capability.[45]

In summary, the Somalia experience in 1995 exposed a lack of continuity between the Pentagon's NLW policy, ROE and legality, doctrine and training. In light of these shortcomings, it is questionable whether the equipment should have been deployed at all, given the dearth of experience and policy direction. Indeed, more than anything else, NLW deployment would appear to have been driven by sensitivity to casualties and difficulties in interpreting appropriate ROE. Nevertheless, in order to address these apparent frailties, NLW doctrine and the Force Continuum have been further developed to articulate their relationships with conflict intensity.

Non-Lethal Weapons Doctrine and the Conflict Intensity Continuum

Following Operation 'United Shield', a significant gap in NLW doctrine became apparent to several US military planners.[46] The first MEF had deployed NLWs without sufficient development of the conceptual framework within which they would be used. Equally, the dynamics of the interaction between peacekeeping troops and crowds had not been fully explored. Furthermore, NLWs were classified under the same ROE as lethal weapons. In effect, this meant that a deadly threat would have to be presented before NLWs could be used. Consequently, the concept of a Conflict Intensity Continuum was developed, complementing the Force Continuum, as illustrated in Figures 2 and 3.

The relationship outlined in Figure 2 reflects the post-cold war security environment, while recognizing the utility and application of military force across a wide range of potential missions. When compared to the 'Force Continuum' in Figure 3 one can see that both concepts are represented in a linear manner. It is argued that these models highlight and address the capability gap between the show of force and the application of lethal force and that NLWs are the ideal means to fill this gap. Furthermore, it is argued

FIGURE 2

THE CONFLICT INTENSITY CONTINUUM

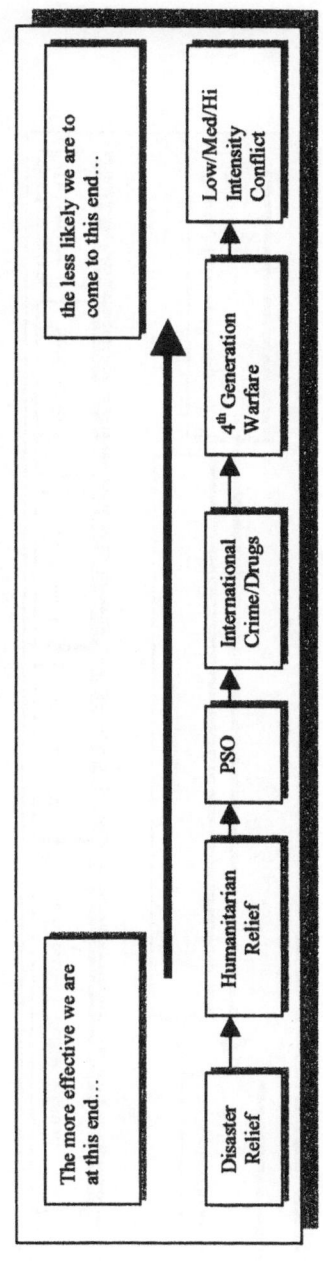

The more effective we are at this end…

Disaster Relief

Humanitarian Relief

PSO

International Crime/Drugs

4th Generation Warfare

Low/Med/Hi Intensity Conflict

the less likely we are to come to this end…

Taken from Lt.-Gen. Anthony Zinni and Col. Gary Ohls, 'No Premium on Killing', *US Naval Institute Proceedings*, Dec. 1996.

FIGURE 3
THE FORCE CONTINUUM

Taken from Lt.-Gen. Anthony Zinni and Col. Gary Ohls, 'No Premium on Killing', US Naval Institute Proceedings, Dec. 1996: 27.

that if NLWs are applied across the wide range of contingency operations at the lower end of the conflict intensity continuum, they provide an effective means of preventing crisis escalation.[47] The argument proposes that such an approach is flexible and reduces the risk of injury to friendly forces and individuals in the target crowd and minimizes collateral damage. If one accepts the assumption that these linear relationships exist, then the interaction between friendly forces and the target crowd is predictable and relates directly to outcome at any point along the continuum. Moreover, there is an assumption that the friendly force has direct control and influence on the conduct of a crisis event such as a demonstration or a riot. However, the crisis dynamic may not follow such a predictable and concise line of interaction.

If the target population/leadership identifies this predictability of response and assumption of control, then it is more, rather than less, likely to be exploited asymmetrically. Indeed, the models are postulated from a western cultural perspective and do not take into account the different social, economic and cultural backgrounds of target populations, which may possess diverse values, goals and perceptions. As highlighted previously, these are key features of the post-cold war security paradigm. Consequently, a target leadership/population could seek to influence a crisis asymmetrically by accelerating events towards the right of the conflict intensity continuum.

Moreover, as discussed earlier, the media could be exploited to influence wider opinion in favour of the target population and to isolate friendly forces in a non-consensual environment. Paradoxically, the risk of such an approach is that the use of lethal force would be more, rather than less likely, despite the availability of NLWs. Thus, the likelihood of injury and damage to property would increase. In effect, NLW deployment could hinder the achievement of friendly policy goals with the potential of greater injurious effect – through the application of lethal force – for both friendly forces and the target population.

Notwithstanding these potential pitfalls, the Force and Conflict Intensity Continuums have become cornerstones of American NLW doctrine in the 'Joint Concept for Non Lethal Weapons'. This concept supports the current American structure for NLW evaluation and doctrinal development.

The Joint Non-Lethal Weapons Directorate and NLW Evaluation

The 'Joint Concept for Non Lethal Weapons' was published in January 1998[48] and arose from 'Joint Vision 2010' and the pursuit of 'the ability to produce a broader range of potential weapons effects' while supporting the US operational concept of 'full dimensional protection'.[49] The concept identifies eight guiding principles for NLW development:

- *Leverage High Technology* – To think creatively, in partnership with industry, in order to exploit new technology for adaptation as a NLW system;

- *Enhance Operations* – Achieve net improvement in readiness, performance and capability;

- *Augment Deadly Force* – NLWs are additional to conventional systems to limit the probability of death for both combatants and non-combatants;

- *Provide a 'Rheostat' Capability* – Deliver varying levels of effect in a flexible, 'rheostatic' manner;

- *Facilitate Expeditionary Operations* – NLW qualities should include mobility, endurance and sustainability;

- *Maintain Policy Acceptability* – In political, social and legal terms;

- *Provide Reversibility in Counter Personnel Effects*

- *Apply across the Range of Military Operations* – While NLWs would be considered most appropriate at the tactical level of warfare, it should be recognized that they may be applied at both strategic and operational levels, potentially incorporated into a scheme of manoeuvre.

On 23 June 1999, the US Joint Chiefs of Staff subsequently issued the Joint Service Memorandum of Agreement for the DOD Non-Lethal Weapons Program.[50] The Memorandum established the Commandant of the USMC as the Executive Agent for co-ordinating and directing NLWs plans and programmes through the Joint Non-Lethal Weapons Directorate (JNLWD). In line with the Clinton administration's 'dual-use' strategy of developing technologies with both civil and military applications, the JNLWD is closely integrated with industry and research agencies. The Executive Agent is responsible to the DOD as the focal point for NLWs research and chairs an Integrated Product Team (IPT) comprising representatives from each service, the DOD and US government transport and industry agencies. The IPT addresses many of the deficiencies highlighted previously in doctrine, training and the acquisition of NLWs and provides an overarching view of the development process within an organized and co-ordinated structure.[51]

JNLWD research and development is directed in five main areas: acoustic weapon systems, entanglements, kinetic weapons, riot control agents and vehicle stoppers.[52] In addition to military research projects, commercial off-the-shelf solutions (COTS) available within the law enforcement community are also being evaluated.[53] Examples include laser flashlights designed to dazzle rather than blind, capture nets, ring airfoil projectiles, electric stun projectiles and pepper spray projectiles. However, there are some potential legal implications in terms of the liability of military personnel and industry in this relationship should a COTS product kill or maim.[54]

To inform the evaluation process, in 1998 Pennsylvania State University was commissioned to set up the Human Effects Advisory Panel (HEAP) under a five-year Marine Corps Research University contract worth $42.5 million.[55] The HEAP mission is to provide independent assessment of human effects issues surrounding NLWs. The first task set was to assess the methodology applied by the US services in evaluating blunt human trauma. The HEAP was critical of current methodologies because the models used were neither validated, nor did they address likely injury modes. To begin with, the HEAP attempted to describe 'non-lethal' in a qualitative manner and a guidance metric – as shown in Figure 4 – was produced:

The distribution curve attempts to depict the effect of an ideal NLW on a population, noting that there will be a fraction at either end of the scale where no effect or permanent injury takes place. However, the JNLWD – while accepting the metric as a descriptive model of the goal of the NLW programme – advised that it should not be used as a 'go/no go test for the acceptability of any specific system'.[56]

Furthermore, the HEAP proposed methods of generating and presenting human data using established bio-statistical techniques and existing models such as clay, gelatine, dummy technology and animal testing. The JNLWD concluded that fielded weapon systems must be accompanied by a comprehensive data package incorporating information on the health effects on users, bystanders and targeted individuals and weapon effectiveness. However, despite this on-going work, there remains the moral issue of disabling an individual, yet reserving the right to apply lethal force. Indeed, at what level of disability could such action be judged as militarily necessary? As Dando observes, balancing the claims of military necessity and inhumanity 'bedevil the whole problem of developing international law in this area'.[57]

The Non-Lethal Weapons Effects Profile, Military Necessity and the Humanity Principle

As Coupland states, if the energy output or effect of a weapon can be measured in terms of disability against time then the line between 'non-lethal temporary effects' and 'lethal permanent effects' has been defined.[58]

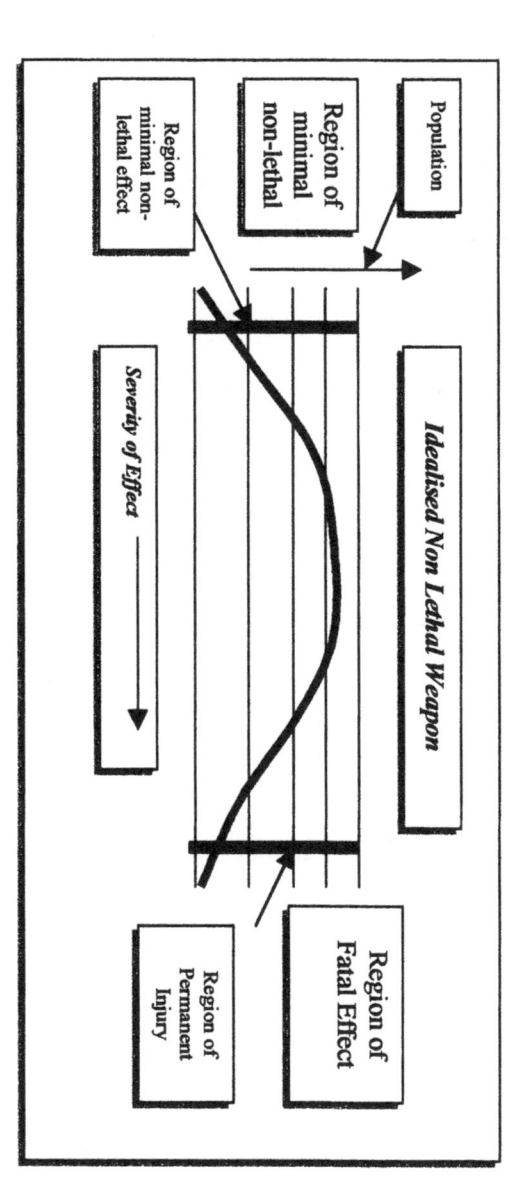

FIGURE 4
IDEALISED NON-LETHAL WEAPON

Idealised NLW Planning Curve, 'HEAP: Blunt Impact Munitions', *JNLWD Reviews* 3rd Quarter 1999; Penn State University (http://www.psu.edu/ur/NEWS/news/marineu.html).

Yet, while providing a model to describe this demarcation, the HEAP distribution curve raises further questions about NLW utility, particularly when complemented by lethal force. By definition, a normal distribution curve can be divided into percentile regions for a given population. Therefore, the curve could be used as a tool to express both the level and percentage disability within a crowd at which it is legally acceptable to apply lethal means.

However, while ROE may provide a military commander the authority to switch between 'non-lethal' and 'lethal' force as deemed appropriate by military necessity, such a decision must be balanced against both proportionality and discrimination under the Law of Armed Conflict. Thus, the application of lethal force above or below a given percentile on the curve could be construed as legal or illegal.

Furthermore, known data on the time duration of a reversible NLW effect could be interpreted in the same manner. Thus, the question is whether a statistical methodology could be developed to provide the basis upon which such actions may be considered humane or inhumane? A complicating factor is the risk that ROE, rather than allowing a flexible response, can become complex and restrictive thereby hindering mission command. Inevitably, an inappropriate response or delays in decision-making will place both military personnel and the target population at greater risk of injury or death. One example is an incident involving a US Marine in Somalia when his colleagues relied on pepper spray to deter an assailant: 'Although the spray worked and the soldier escaped unharmed, the Somali had attempted to stab one soldier four times before he was subdued with the spray.'[59]

In light of the emphasis placed upon aggravation of suffering, the use of conventional weapons to apply lethal force following disablement with NLWs could be interpreted as 'contrary to the laws of humanity'. For NLW opponents who question this particular aspect of their utility, the balance of military necessity and humanity has resonance given the range of NLW development programmes.[60] Yet, it also highlights a weakness in international law – that banning the individual technology alone does not suffice. The key principle is to address the 'effect' of the technology in quantifiable terms.[61]

Analysts such as Krepon[62] have argued that there is a powerful dynamic between weapon designers, the military and the government which emphasizes the beneficial claims made for new weapon systems but to the detriment of the humanitarian dimension of policy-making. The implication for NLW assessment is that aspects of their evaluation and development may be hidden from wider public scrutiny. Krepon suggests four guidelines to help inform the process of determining whether weapons contravene humanitarian values:

- Obtain early political input to the development process;
- Evaluate collateral damage as the weapon is developed;
- Prepare guidelines for anti-personnel weapons use prior to engagement;
- Conduct field checks once weapons are operational.

This aspect of weapons development must be balanced against the competing interests of the military-industrial research complex and the drive to maintain influence and a strong research infrastructure.[63] However, the judgement as to whether the use of a weapon can be deemed moral or militarily necessary must be based on the manufacturer's and the end-user's understanding of a given weapon system – in terms of its design and the predictability, and therefore the foreseeability, of its human effects.

The Legality of Non-Lethal Weapon Effects

During the development of a weapon system the cornerstone of the evaluation process involves observation of its efficacy by simulation and laboratory or controlled field experiment, the results of which facilitate modification and refinement. Thus, there is a cycle of activities principally concerned with refining the design parameters of a weapon system.

However, the acid test of a weapon's true efficacy is its performance in the hands of the end-user. This leads to a second cycle of assessment and evaluation, which involves observation of the weapons' human effects in the field. Historically, this task has fallen to both military and civilian health professionals. The medical evidence for this process is legion with examples including mustard agent, dum-dum bullets and anti-personnel mines. It is this second cycle that protects the interests of the potential victims of a weapon system by providing data and evidence on its effects to the wider political, military, medical and legal communities. In essence, this second cycle informs both policy and law.

The Solferino Cycle and Non-Lethal Weapons

Henry Dunant was the first to initiate this second cycle of assessment when he observed the effects of munitions on soldiers at the Battle of Solferino in 1859. Dunant's writings on the realities of warfare, in *A Memory of Solferino*, and his subsequent involvement in the formation of the International Committee of the Red Cross (ICRC), provided the catalyst for the First Geneva Convention in 1863.[64] Over the following 100 years, chemical weapons, exploding bullets, landmines and lasers have all been assessed in a similar manner leading to the term 'the Solferino Cycle'.[65]

The cycle is inextricably linked to the observation of the human effects of weapons, developed as a result of the availability of new technology, which produce injuries perceived as either abhorrent or inhumane. Data and statistical evidence collected on weapon effects have been used to inform and, more importantly, to shape international law. However, the final loop in the Solferino Cycle – the communication of the evidence of a weapon's effect profile to the wider community – has not always been timely, well articulated or, indeed, favourably received by policy-makers. Anti-personnel mines and the tortuous route to the Ottawa Treaty are a case in point, despite extensive evidence of their indiscriminate injury patterns.[66]

As noted previously, there is evidence of a wide pattern of potential human effects related to both the design and function of NLWs such as kinetic rounds.[67] It is possible that such effects may be foreseeable at the design phase. This potentiality behoves weapon designers to consider this issue at all stages of the evaluation process and to assess the discrimination properties of the weapon design; the range of weapons that can be developed and deployed is not unlimited.[68]

However, the potential human impact of a NLW may not be predictable. Indeed, the injury patterns of a NLW may fall outside current medical experience and, therefore, remain unforeseen by the designer and end-user. To examine this point further, it is worthwhile considering the broad classes of NLWs against current law and the principle of 'superfluous injury and unnecessary suffering'.

Non-Lethal Weapons and Superfluous Injury or Unnecessary Suffering: The Superfluous Injury or Unnecessary Suffering (SIrUS) Project

Weapons can be assessed under two broad legal areas including the Law of Armed Conflict and those Principles Governing Weapons founded on Judeo-Christian law.[69] These form the basis upon which states can articulate and adopt specific prohibitions and restrictions on particular weapon systems. Proportionality, military necessity and humanity are addressed under the Law of Armed Conflict. However, history has shown that these are the principles open to the widest interpretation within the international community. Three additional principles – specifically governing weapons – have also been articulated, namely 'unnecessary suffering', 'discrimination' and 'treachery or perfidy'.

The principle of 'unnecessary suffering' provides a legal and ethical basis against which the utility of a weapon can be measured. Indeed, it has acted as the stimulus for specific bans on weapons such as dum-dum bullets.[70]

The 'discrimination' principle prohibits the use of methods or means of warfare that cannot be directed against a specific military objective and, therefore, are of a nature to strike military and civilian objectives without distinction.[71] Anti-personnel mines, as discussed earlier, have been addressed under this category.

The principle of 'treachery' or 'perfidy' prohibits certain perfidious uses of weapons and may prohibit weapons that are inherently perfidious.[72] However the limits of this principle are far from certain. This was highlighted in the Scientific and Technological Options Assessment Committee of the European Parliament commissioned study entitled 'An Appraisal of the Technology of Control'. This study documented cases where police forces used NLWs as instruments of torture or in a cruel or inhumane manner.[73]

From these broad legal areas, numerous weapon declarations and conventions have been derived including the 1993 Chemical Weapons Convention, the 1972 Bacteriological and Toxin Weapons Convention and the 1980 UN Inhumane Weapons Convention.[74] It could be argued that the multiplicity of conventions facilitates a wide variety of interpretations of the purpose and utility of a weapon. This is especially the case when considering NLWs. For example, are bacteriological agents employed to degrade petroleum products as 'vehicle stoppers' discriminate? Furthermore, could they have wider health effects? Equally, do psychotropic agents used as a 'calmative' riot control means fulfil the same criteria? Thus, which law or arms control convention applies?

Clearly, purpose, utility and intent may be interpreted from a variety of viewpoints under existing measures if there are no objective, quantitative means available through which to make a calculated assessment of a weapon. Moreover, the competing interests of end-users, governments or arms control implementation bodies further confound the process. As a consequence, it may be difficult to reach a consensus on a particular weapon's characteristics, which can lead to a failure to, or at least a delay in, establishing arms control agreements.

One method of simplifying this dilemma would be to develop agreed criteria against which any generic weapon system could be assessed. The ICRC has proposed such a methodology. In 1996, the ICRC conducted a wide-ranging symposium on the medical profession and the effects of weapons.[75] The aim was to establish the key role that health professionals play in collecting and providing data on weapon effects in order to move towards an evidence-based system of integrating legal, military and political concerns on current and future weapons development – the Solferino Cycle revisited. The following year, the ICRC proposed the formation of the Superfluous Injury or Unnecessary Suffering (SIrUS) Project[76] drawing upon legal, medical and military experience and opinion.

The basis for SIrUS was the ICRC database of information on 26,636 war-wounded individuals admitted to hospitals during conflicts. From this database an objective assessment of clinical outcomes – in terms of mortality and morbidity – against different classes of conventional weapons could be established. Variables in the analysis included the proportion of large wounds, relative proportions of central and limb injuries, duration of hospital stay, the number of operations required, the requirement and volume of blood transfusions and the extent of severe and permanent disability in survivors. This data was correlated with evidence from military publications.

From this summation the distinction between the effects of conventional weapons and all other weapons could be described, thus determining what is not superfluous injury or unnecessary suffering. Furthermore, the project team established that particular weapon effects had not been observed as the result of armed conflict in the last 50 years. These include:[76]

- Disease other than that resulting from physical trauma from explosions or projectiles;
- Abnormal physiological or psychological states (other than conditions such as Post-Traumatic Stress Disorder);
- Permanent disability specific to the kind of weapon (with the exception of point detonated anti-personnel mines);
- Disfigurement specific to weapon type;
- Inevitable or virtually inevitable death in the field or high hospital mortality level;
- Grade 3 wounds among those who survive to hospital;
- Effects for which there is no well recognized and proven medical treatment that can be applied in a well-equipped field hospital.

The SIrUS Project committee has proposed that superfluous injury and unnecessary suffering is determined by the design dependent, foreseeable human effects of weapons. Thus, superfluous injury and suffering falls under one or more of the following criteria:

- A specific disease, abnormal physiological state, psychological state, permanent disability or disfigurement;
- Field mortality of more than 25 per cent or hospital mortality of more than five per cent;
- Grade 3 wounds as measured by the Red Cross Wound Classification;
- Effects for which there is no well-recognized and proven treatment.

The ICRC has made two proposals derived from the above. First, when assessing the legality of a weapon, governments should establish whether it would cause any of the effects listed above by function of its design. If so, governments should then weigh the military utility of the weapon against these effects to establish whether the same purpose could be reasonably achieved by other lawful means that do not produce such effects. Second, government should make efforts to build a common understanding of the norms to be applied in reviewing new weapons and to promote transparency in the conduct and results of such reviews.[76]

These proposals may be viewed as a Utopian ideal given the complexities generated by the national and industrial interests highlighted earlier. However, they do provide an evidence-based method by which to assess a weapon. Paradoxically, if these proposals are adopted, they could serve to protect national and industrial interests in the long term because of the potential legal, political and financial pitfalls for both government and industry of deploying a weapon the subsequent use of which ultimately results in charges of superfluous injury and unnecessary suffering. This point was highlighted in an appraisal of acoustic weapon evaluation. Unless more attention was paid to the bio-effects of NLWS, it was argued that, 'Expensive hardware could be developed that was operationally useless, prohibited by policy or had extremely variable effects'.[77]

If the ICRC criteria and proposals are adopted by the international community, the implications for the development of NLWs will be wide-ranging. Each of the criteria can be applied to specific classes of NLWs – those technologies currently under development or consideration are listed in Figure 5.

FIGURE 5
NLW TECHNOLOGIES IN DEVELOPMENT / CONSIDERATION

Lasers	Optical Munitions	Acoustics
Hand-held	Broad beam isotropic	Acoustic Beam
Light-infantry	Directed optical (dazzle)	Acoustic Bullet
Pulsed Chemical	Infrasound	
	Vortex Ring Generation	
Electro-magnetic Pulse	Chemical & Biological	Kinetic Rounds
Systems	Foam and sticky materials	Grenade launched foam
Grenade launched	Anti-material Compounds	Rubber Bullets
electrical 'sting-ball'	Fuel degradants	Bean bags
Super corrosives	Liquid Metal	Calmative Agents
	Embrittlement	Di-methyl Sulphoxide
		Hydrogen Sulphide

For example, directed energy weapons such as acoustic, infrasound or microwave systems – designed to disable or disorientate by targeting their actions on a specific system such as the gastro-intestinal tract or central nervous system – will fall under criterion (a). Equally, anti-material or anti-personnel chemical and bacteriological agents could fall under both (a) and (d). Dependent on their design parameters, however, kinetic rounds may fall outside all of the criteria.

In short, the ICRC criteria provide a method of objective assessment based upon data analysis, which can inform the weapon evaluation process without recourse to new laws, conventions or treaties. The criteria avoid the doctrinal confusion of establishing the differences between NLWs and conventional weapons. Moreover, the ICRC methodology has moved towards a determination of the principle of superfluous injury and unnecessary suffering which can be measured against a generic weapon system.

In addition, the proposals offer the opportunity to design out illegal features or simply to abandon weapon development before its effects become evident in the field. A window of opportunity exists that may ensure the mistakes of the past are not repeated in the future.

Finally, incorporation of the proposals into existing law could help to avoid the following vision of future conflict: 'Rather than sutured wounds, skin grafts or amputations, will the soldiers who have survived battlefields of the future return home with psychoses, epilepsy and blindness inflicted by weapons designed to do exactly that?'[78]

Conclusion

NLWs are a product of the rapidly changing security environment of the post-cold war world and revisionist thinking in political and military doctrine for intervention operations. They are also a product of the genuine belief among NLW advocates that there is a 'better way' to apply force. The definition 'non-lethal', however, is fraught with expectations and connotations of NLW utility and effectiveness. This chapter has not criticized the conceptual thinking behind the utility of NLWs *per se*. Rather, it has attempted to highlight areas that require further development if NLWs are to be considered truly effective. At defence policy level, NLW procurement within an already stretched defence budget may have repercussions for long-term costing and planning.[79]

Furthermore, NLW concepts may create undue complexity in the preparation and execution of military operations, in addition to a greater training burden. Doctrine must be further developed to address paradoxes in areas such as ROE and the potential asymmetric exploitation of NLWs by target populations/leaderships. Most importantly, the utility of NLWs in relation to negotiation, communication and confidence building measures must be articulated within a broader PSO doctrine. A weapon system developed simply to replace or circumvent these approaches could ultimately place the user on the ground, and the target population, at significantly greater risk of injury or death due to a loss of consensus and support.

Equally important is an appreciation of the different cultural perspectives of both the user and the target population. Simply relying upon NLWs in complex emergencies cannot replace the difficult task of understanding and addressing the underlying factors associated with managing a crisis or a conflict; to do so is to distance oneself from the problem. One can postulate that NLWs are an expression of western casualty aversion and increasing concern for friendly force protection.

Yet, as mentioned earlier, total mortality and morbidity has remained relatively constant since the Second World War. The paradox lies in the fact that while casualty rates among friendly forces may be low, casualty sustainment – almost by default – has moved into the civilian non-combatant population. Furthermore, there is a growing body of medical evidence – from civil, military and law enforcement sources – that the use of NLWs in precisely the type of urban environment highlighted by NLW advocates will lead to greater mortality and morbidity. Thus, the question arises whether NLWs can reduce or accelerate this disturbing trend?

Notwithstanding the issue of tactical utility, the broader operational and strategic effects, both direct and indirect, on a population of using NLWs such as carbon ribbon and computer viruses must be assessed and balanced against both political and military necessity. Moreover, there is the difficult question of human rights and the perfidious use of NLWs. NLWs are weapons like any other – they are intended to cause harm. The difficulty facing politicians, the military and scientists is how to provide a quantitative measure of the effects and benefits of using NLWs over other means.

Each constituency has a responsibility to evaluate a weapon system at all stages of development and to determine its legality under the Law of Armed Conflict and the Principles Governing Weapons. History is littered with successes and potential failures in this area; examples include dum-dum bullet technology and delays in addressing the effects of blinding laser systems and anti-personnel mines.

Health professionals have historically provided the evidence of a weapon's short- and long-term effects. For medical personnel in the military, there is no conflict of interest in carrying out this activity; it is a professional responsibility to highlight new developments in weapons technology that by their design or effect may have significant health impacts for combatants and non-combatants. However, failure to articulate the data, failure on the part of politicians to appreciate the scale of effect and failure to achieve a consensus of political opinion have all conspired in the past to delay or block effective arms control measures. Moreover, a wide variety of conventions and treaties, combined with their flexible interpretation and application, compound the weapons evaluation cycle.

Clearly, the wide array of NLWs technologies available for current and future employment present a range of challenges in terms of implementing effective arms control measures. Direct-energy systems, kinetic rounds and chemical and bacteriological agents could all be addressed under existing treaties or conventions. However, for the reasons listed earlier, this process is inherently difficult without a fundamental change of approach by governments to arms control agreements. Without consistency and uniformity, there are grave risks of proliferation or national 'opt outs'. Security issues and the competing interests of industry will further compound this problem.

A more realistic approach, and one that may guarantee success, would be to examine NLW technologies in the same manner as conventional weapons. The ICRC criteria provide an evidence-based assessment tool that can be applied to any current or future weapon system. Thus, endorsement of the criteria, and their incorporation into existing law, could provide the means to regulate and validate the effects of a weapon system, rather than the difficult and tortuous process of evaluating the individual technology.

Further to the ICRC criteria, the R²IPE acronym is proposed as a means of highlighting the key areas that are peculiar to NLWs and which should be considered in the weapon development and acquisition process:

R² – Reversibility of Effect and the 'Rheostat' Capability

The reversibility of effect and the capability to provide 'rheostat' capabilities in the application of force are unique to NLWs and central to current US concepts. Clear definition of the duration of NLWs efficacy, residual effects and whether the NLWs effect is self-limiting, or requires active intervention or treatment, is required. Furthermore, clear guidelines should be established as to when the application of lethal force is legally acceptable and this should be articulated fully in ROE.

I – Information

The evaluation of NLWs should be an open process to allow unequivocal and independent assessment of a system's effects. Clearly, this will require a significant culture change with regard to issues of security and the fear that countermeasures could be developed. Nevertheless, to date it has been restrictions on the release of information and subsequent analysis of data that has limited the effectiveness of weapon evaluation and the wider scrutiny of their utility. The costs to industry and government must be balanced against the human cost in lives and the financial cost of treating the injured or maimed.

P – Policy

NLW policy should be clearly linked to all available data. This will provide a solid foundation upon which NLW doctrine can be developed. Thus, avoiding the confusion and misperceptions that occurred during Operation 'United Shield'.

E – Effect

The efficacy a NLW system should be fully assessed against the ICRC criteria to establish whether the weapon contravenes the principle of superfluous injury or unnecessary suffering. This is the key principle in NLW evaluation.

The ICRC criteria, in addition to those listed above, provide a framework around which to develop a method to appraise NLW systems without resort to specific arms control measures. For the first time there appears to be an opportunity to control weapon technologies before their effects are seen in the field. All that remains is the question of whether the international will exists to complete the NLW 'Solferino Cycle'.

Notes

1. Steele M. Address at the Non-Lethal Defence III Conference, Johns Hopkins University, 25 Feb. 1998, www.jnlwd.usmc.mil/default2.htm.
2. Reitinger K. New Tools for NewJobs. *US Naval Institute: Proceedings* April 1998: 38.
3. Heal C. Nonlethal Technology and the Way We Think of "Force". *Marine Corps Gazette* Jan. 1997; **81** (1): 27.
4. Roland-Price A. Can Future War be Non-Lethal? *British Army Review* Aug. 1996; **113**: 22.
5. NATO AGARD Report. *Minimising Collateral Damage during Peace Support Operations No.8.* May 1997.
6. Note 2: 38.
7. Note 3: 27.
8. Garfield M, Neugut A. Epidemiological Analysis of Warfare: A Historical Review. *JAMA* 1991; **266**: 688–92.
9. Ibid.: 692.
10. Bellamy RF. The Medical Effects of Conventional Weapons. *World Journal of Surgery* 1992; **16**: 888–92.
11. Meilinger PS. Ten Propositions Regarding Air Power. *Air Power* Spring 1996; **10** (1): 50, 52–71.
12. Coupland R, Meddings D. Mortality Associated with Use of Weapons in Armed Conflicts, Wartime Atrocities, and Civilian Mass Shootings: Literature Review. *BMJ* 1999; **319**: 409.
13. Hiss J, Hellman F, Kahana T. Rubber and Plastic Ammunition Lethal Injuries: The Israeli Experience. *Medicine, Science and the Law* 1997; **37** (Part 2): 139–44.
14. Bailey W. Less-than-Lethal Weapons and Police-Citizen Killings in US Urban Areas. *Crime and Delinquency* 1996; **42** (4): 553.
15. DARPA Tech 99, June 1999, Defence Advanced Research Projects Agency, http://www.darpa.mil/.
16. Protocol on Blinding Laser Weapons (Protocol IV – addition to 1980 UN Convention on Certain Conventional Weapons) 13 Oct. 1995; Ottawa Treaty and Convention on the Prohibition of the Use, Stockpiling, Production and Transfer of Anti-Personnel mines and on their Destruction, 18 Sept. 1997.
17. Coupland R. The Effect of Weapons: Defining Superfluous Injury and Unnecessary Suffering. *Med Glob Surv* 1996; **3** (3-A1), www.ippnw.org/MGS/.
18. Geneva Conventions of 1949, Additional Protocol I (1977), Article 35(2).
19. Hague Regulations, Article 23(e).
20. International Committee of the Red Cross (ICRC). *Towards a determination of which weapons cause 'superfluous injury or unnecessary suffering*, 10 Nov. 1997, www.icrc.org/.
21. Lewer N. *Research Report No 1.* Centre for Conflict Resolution, University of Bradford,

Nov. 1997, www.brad.ac.uk/acad/nlw/cendera1.html.
22. Karnow S. *Vietnam: A History*. London: Pimlico, 1994: 33.
23. Ibid.: 33.
24. Lewer N, Schofield S. *Non-Lethal Weapons: A Fatal Attraction?* London: Zed Books, 1997: 6.
25. See Lamb C. *Non Lethal Weapons Policy*. US Department of Defence Directive 1 Jan. 1995; US DoD Directive No 3000.3, *Policy for Non Lethal Weapons*, 6 July 1996, para. C 2(a), www.jnlwd.usmc.mil/default2.htm.
26. NATO Statement on Non-Lethal Weapons. *Disarmament Diplomacy* 1999; **40**.
27. Metz S. *Discussion Paper: Revolutionary Perspectives on the Revolution in Military Affairs*. Revolution in Military Affairs Conference, Australian Defence Studies Centre, 27–28 Feb. 1996: 2.
28. Lt.-Gen. A Zinni (USMC) was US Force Commander for Operation 'United Shield' in Somalia.
29. Zinni A, Ohls G. No Premium on Killing. *US Naval Institute: Proceedings* Dec. 1996: 27.
30. Sun Tzu. *The Art of War*. Oxford: Oxford University Press, 1971: 77.
31. Ibid.: 78–9.
32. Note 24: 116.
33. Stanton M. What Price Sticky Foam? *Parameters* 1996; **27** (3): 66–7.
34. Peters R. The New Warrior Class. *Parameters* 1994; **25** (2): 25.
35. Leonhard RR. *The Principles of War for the Information Age*. Novato, CA: Presidio, 1998: 24.
36. Note 33: 66.
37. Note 29: 27.
38. Anderson G. There's a Better Way. *Armed Forces Journal International* July 1996: 15.
39. Lorenz FM. Non-Lethal Force: The Slippery Slope to War? *Parameters* 1996; **27** (3): 53.
40. Note 39: 52.
41. Note 29: 27.
42. Note 39: 55.
43. Note 39: 56.
44. Note 33: 64–5.
45. Note 39: 62.
46. Becker JB, Heal C. Less than Lethal Force: Doctrine Must Lead the Technology Rush. *Jane's International Defence Review*, Feb. 1996.
47. Note 29: 27.
48. *Joint Concept for Non-lethal Weapons*, Jan. 1998, Joint Non-Lethal Weapons Directorate, www.jnlwd.usmc.mil/default2.htm.
49. The ability to accomplish assigned missions while simultaneously reducing the adverse effects of military operations, especially collateral damage.
50. *Joint Service Memorandum of Agreement for DOD NLW Programs*, 23 June 1999, Joint Non-Lethal Weapons Directorate, www.jnlwd.usmc.ind/default2.htm.
51. Joint NLW Integrated Product Team Charter, Annex A, p A-4, Paras.4.1–4.4, 23 June 1999; Joint NLW Directorate Charter, Annex C, p C-2 Figure 1, 23 June 1999.
52. JNLWD Programme, www.jnlwd.usmc.mil.
53. Siuru B. Developments for the military and law enforcement. *Corrections Technology & Management* March/April 1999: 22. National Law Enforcement and Corrections Technology Center, nlectc.org/specialannouncements/siuru.html.
54. Pohling-Brown P. Non-Lethal Weapons: Who Pays if they Kill? *Jane's Defence Contracts* Nov. 1996: 4–5.
55. Transcript taken from Penn State University website, www.psu.edu/ur/NEWS/news/marineu.thml; HEAP: Blunt impact munitions. *Joint Non Lethal Weapon Directorate (JNLWD) Review*, 3rd Quarter 1999, www.marcorsyscom.usmc.mil/jnlwd/documents.
56. Ibid.: 3.
57. Dando M. *A New Form of Warfare: The Rise of Non-Lethal Weapons*. London: Brassey's, 1996: 30.
58. Coupland R. Non-Lethal Weapons: Precipitating a New Arms Race. *BMJ* 1997; **315**: 72.
59. Dworken JT. Rules of Engagement: Lessons Learned from Restore Hope. *Military Review* 1994; **74** (9): 31.
60. Note 58: 72.

61. Coupland R. The Effect of Weapons on Health. *Lancet* 1996; **347**: 450.
62. Cited in Lewer and Schofield, Note 24: 82.
63. Cited in Lewer and Schofield, Note 24: 134.
64. Coupland R. The effects of weapons and the solferino cycle. *BMJ* 1999; **319**: 864.
65. Ibid.: 865.
66. Hewish M, Pengelly R. In Search of a Successor to the Anti-Personnel Landmine. *Jane's International Defence Review* 1998; **19** (3): 30.
67. Note 13: 144.
68. 1977 Protocol 1, Additional to the Geneva Conventions of 1949.
69. Note 24: 83.
70. The Declaration of St Petersburg, 1868, The Hague Regulations, Article 23(e) and the Geneva Conventions 1949 (Additional Protocol 1 1977, Article 35[2]).
71. Geneva Conventions of 1949 (Additional Protocol 1 1977, Article 51[4]).
72. The Hague Regulations, Article 23(b).
73. Transcript available, www.jya.com/STOA-atpc.html.
74. Note 24: 86.
75. *The Medical Profession and the Effects of Weapons*, International Committee of the Red Cross (ICRC) Symposium, 8 March 1996, www.icrc.org.
76. *Towards a determination of which weapons cause superfluous injury or unnecessary suffering*. International Committee of the Red Cross (ICRC) Symposium, 10 Nov. 1997, www.icrc.org/.
77. Murphy M. *Biological Effects of Non-Lethal Weapons: Issues and Solutions*. Paper delivered at NLD III Conference, John Hopkins Applied Physics Laboratory, 25 Feb. 1998.
78. Spinney L. A Fate Worse Than Death. *New Sci*, 18 Oct. 1997: 26.
79. Lewer N. *Non-Lethal Weapons Research Project Paper 2*, June 1998, Centre for Conflict Resolution, University of Bradford, www.brad.ac.uk/acad/nlw/censdera.html.

Future Incapacitating Chemical Agents: The Impact of Genomics

MALCOLM DANDO

On 9 February 1998 the then UK Defence Secretary, George Robertson, announced to the House of Commons that:

> the House will wish to know that I am today making available new information on Iraq's chemical weapons capability at the time of the Gulf war. That concerns recently received intelligence that Iraq may have possessed large quantities of a chemical weapons agent known as Agent 15 since the 1980s ... [1]

An accompanying Ministry of Defence (MoD) document released at the same time stated that:

> Agent 15 is one of a large group of chemicals called glycollates (esters of glycollic acid). The best known is usually referred to by the initials BZ. The physiological effects of these compounds are typical of anticholinergic agents, which block cholinergic nerve transmission in the central and peripheral nervous system.[2]

As the document then explained, these glycollates are mental incapacitants:

> we believe that the immediate effects of Agent 15 would include: dilated pupils, flushed faces, dry mouth, tachycardia, increase in skin and body temperature, weakness, dizziness, disorientation, visual hallucinations, confusions, loss of time sense, loss of coordination and stupor.

The UK had good reason for being able to specify these effects; BZ had been weaponized by the United States in the early years of the cold war[3] and the UK had also thoroughly investigated such chemicals.[4] At the time of Robertson's statement, a US Congressional Research Service document asserted that about 100 milligrams of Agent 15 in aerosolized form would be enough to incapacitate someone and that the effects would begin after about 30 minutes and could last for several days.[5]

Against that background, the UK press had a field day with headlines about Iraq's horror weapons[6] and 'zombie gas'.[7] However, the *Daily Telegraph* pointed out that 'Agent 15 is outlawed under the internationally agreed Chemical Weapons Convention'.[8] It would perhaps appear, then, that our concerns over this issue should be limited, since Iraq was acting as a 'rogue state' outside of international law. But is this interpretation of the law universally held, and if not, what might be the consequences in an era of very rapid developments in our understanding of the nervous system?

Incapacitants and the Chemical Weapons Convention Prohibitions

On the understanding of the Chemical Weapons Convention (CWC) that I would wish everyone to hold,[9] the *Daily Telegraph* was quite correct. However, as we must recognize, there is at least a level of ambiguity, and perhaps much more, on this issue in the convention. Indeed, the *Chemical Weapons Convention Bulletin* devoted a major editorial[10] to the problem under the telling title 'New Technologies and the Loophole in the Convention'. The editorial graphically describes the stakes involved:

> The Chemical Weapons Convention in no way limits use of tear gas or other temporarily disabling chemicals by police forces for purposes of domestic riot control. *But the language used to exempt other law-enforcement purposes has created ambiguity in the heart of the Convention* ... [Emphasis added].

In particular, the editorial notes:

> What is at stake is the ability of the treaty regime to withstand technical change. *For new chemical agents and technologies have begun to emerge whose attractions for weapons purposes may eventually drive them through the loophole which the ambiguity has created* [Emphasis added].

TABLE 1
THE LAW ENFORCEMENT LOOPHOLE

- Article II deals with Definitions and Criteria.

- Article II.1(a) is the statement of the general purpose criterion showing that the convention applies to all chemicals except where they are intended for purposes not prohibited.

- Article II.9(d) states that amongst the purposes not prohibited is 'Law enforcement including domestic riot control purposes'.

- Article II.7 defines a riot control agent as 'Any chemical not listed in a Schedule, which can produce rapidly in humans sensory irritation or disabling physical effects which disappear within a short time following termination of exposure'.

- No definition is offered for what chemicals are permitted for law enforcement other than that Schedule 1 chemicals may not be used.

Editorial. New Technologies and the Loophole in the Convention. *Chemical Weapons Convention Bulletin*, March 1994; 23: 1–2.

The editorial then reviewed how the loophole might be seen to have come about. The argument is set out briefly in Table 1.

Given that there is no definition of law enforcement (which is clearly a larger concept than domestic riot control in Article II.9[d]), nor any clear definition of law enforcement chemicals other than that Schedule 1 chemicals may not be used, the question naturally arises:[10]

> Is the Convention really to be read as allowing any non-Schedule-1 toxic chemical or precursor to be developed, produced, weaponised, stockpiled or traded, so long as it is said to be for 'law enforcement purposes'?

The editorial continued:

> The identity of chemicals which state parties hold for riot-control purposes will have to be disclosed in the national declarations under Article III ... [But] ... For chemicals intended for law enforcement purposes other than domestic riot control, there is no provision for any such transparency. *The Convention does not even require disclosure of their chemical names. Their identity, as well as that of munitions and devices for using them, may all be kept secret ...* [Emphasis added].

So it is not clear that the CWC effectively prevents the future development of new chemicals for uses beyond domestic riot control. How did such an ambiguity come about?

It seems that it did not come about by accident for the *Chemical Weapons Convention Bulletin* editorial also stated that:

> Some, by no means a majority, of the negotiating states wished to protect possible applications of disabling chemicals that would either go beyond, or might be criticized as going beyond, applications hitherto customary in the hands of domestic police forces.

It has to be presumed that the countries wishing to protect such applications included the United States since an expert witness told Congress in 1992 that:[11]

> There is no disagreement about allowing the use of certain relatively nonlethal chemicals for purposes of maintaining domestic law and order ... but the present negotiating position of the United States goes far beyond this agreed provision allowing domestic use.

Indeed, this appeared to have been a long-term aim of the United States for the witness continued:

> Starting with its draft treaty of April 1984, *the US position has been to insert a clause in the article on definitions and criteria that would exempt from all provisions of the convention all uses of toxic*

chemicals which are not supertoxic, lethal, or other lethal chemicals,
and which are used by a party for domestic law enforcement and riot
control purposes [Emphasis added].

This exemption in such a blanket form was not agreed, but sufficient
ambiguity to create a grey area certainly was.

Article 1(5) of the CWC, of course, prohibits parties from using riot
control agents as a 'method of warfare', but again this phrase is not defined in
the convention. Moreover, it is a matter of record that President Clinton, in
transmitting the CWC to the Senate, indicated that the US position was that:[12]

> The CWC applies only to the use of RCAs [riot control agents] in
> international or internal armed conflict. Other peacetime uses of
> RCAs such as normal peacekeeping operations, law enforcement
> operations, humanitarian and disaster relief operations, counter-
> terrorism and hostage rescue operations, and non-combatant rescue
> operations conducted outside such conflicts are unaffected by the
> Convention.

The Senate duly included such exemptions in its list of reservations when
ratifying the CWC. The danger is not that new 'law enforcement' chemicals
will be immediately used in all-out warfare, but rather that research and
development and creeping usage will take place within the grey areas
opened up until rich industrial countries see real military advantages in
going down that path.[13]

One particular problem is that, since the end of the cold war, advanced
industrialized countries have committed their forces to many operations
other than war.[14] Often this has involved troops in complex operations
where there were seen to be advantages in the use of non-lethal weapons –
including non-lethal chemical weapons.[15] It is not unreasonable to argue
that – given the military involvement in such operations and the 'maturing'
of non-lethal technologies – this will be a growth area in the US and NATO
countries in the coming years.[16] These developments should not be
underestimated. One US Marine Colonel, also the head of the US Joint
Non-Lethal Weapons Directorate, was reported as stating, 'I would like a
magic dust that would put everyone in a building to sleep, combatants and
non-combatants'.[17]

The Advance of Neuroscience

In its overview of work carried out on glycollates and related compounds
in the UK during the 1960s and 1970s, the MoD noted that:[4]

> The incapacitating potential of anticholinergic drugs was also
> investigated. Comparable potency ratios obtained with these
> compounds in mouse ... and cat studies suggested that the changes

observed in the cat EEG might be mediated by central muscarinic receptors. *These findings supported the suggestion of Downing that the psychotomimetic activity of glycollates was mediated by actions in central muscarinic acetylcholine receptors* ... [Emphasis added].

However, the report went on to state that:

The relationship between the antimuscarinic potency of glycollate compounds and the changes in behaviour are key to an understanding of the mechanisms by which these compounds produce delirium, hallucinations and confusion ... [However,] ... Attempts to establish whether their central actions are a product of interactions at muscarinic receptors were made, but *failed to relate central muscarinic potency to behavioural modifications* ... [Emphasis added].

Whilst other studies of disruption of a conditioned avoidance response were more successful, the difficulties encountered by the investigators are understandable, given the limited knowledge of receptors in the nervous system at that time.

In his keynote contribution to the *Annual Review of Pharmacology and Toxicology* in 2000, Burgen pointed out that:

By 1962, it was becoming clear that the receptor concept was the basis of most, if not all, drug action and that most receptors were likely to be the site of action of endogenous regulators, mostly still unknown and whose role was yet to be defined.[18]

This was a very significant stage in the history of neuroscience,[19] but as Burgen put it, '[s]o far only the tip of the iceberg had been uncovered'. He suggested, in particular, that the future of pharmacology would be concerned with four key problems (Table 2).

TABLE 2
KEY PROBLEMS OF PHARMACOLOGY

1. Defining the receptors, including isolating and characterising them;

2. Finding the endogenous regulators;

3. Developing drug analogues; and

4. Trying to understand the basis of the specificity of drugs for receptors.

ASV Burgen. Targets of Drug Action. *Annu Rev Pharmacol Toxicol* 2000; **40**: 1–16.

In Burgen's view, recent work has yielded much information on receptors and he suggests that:

> It is now clear that the existence of receptor subtypes is a general condition that offers a prospect of finer discrimination, including differentiation of the effector systems to which they are coupled and which could lead to more selective action of drugs.

The difficulties for the scientists working on muscarinic receptors in the 1960s can be gauged from a recent meeting report which began:[20]

> When the *First International Symposium on Subtypes of Muscarinic Receptors* was held in 1983, a key purpose of the meeting was to promulgate the novel concept that there was more than one subtype of muscarinic actetylcholine receptor (mAChR).

The report went on to state that:

> The *Ninth International Symposium on Subtypes of Muscarinic Receptors* was a particularly historic occasion because this was the first meeting at which descriptions of the phenotypes of mice lacking each of the five mAChR genes were available.

Given that genomics is central to our understanding of modern biology;[21] it is not surprising to find that the breakthrough in genomics interacts with advances in other areas such as neuroscience to produce accelerated scientific and technological developments. Whilst it must not be argued that everything about our behaviour can be reduced to genomics,[22] there is no doubt that enormous and rapid changes in our understanding of nervous systems have come about over the last decade.

Neurotransmitters and Neuroreceptors

Forty years ago, when the Ministry of Defence was working on glycollates, the central dogma was that each neuron used one particular chemical substance to influence the behaviour of other neurons to which it was connected. These substances were thought to be small molecules (such as acetylcholine) and very few were known. This idea, first suggested by Dale, is now known to be incorrect. There are many examples in which a neuron is known to release both a 'classical' transmitter such as acetylcholine *and* a neuropeptide at a chemical synapse onto another neuron.[23]

So we know now that chemical transmission at synapses in the nervous system is much more complex than was imagined 40 years ago. Not only are there many more transmitters, some of which are co-located in the same neuron, but there are many more known receptors and receptor subtypes.[24] There are also good, and increasingly better, data available on the structure of these receptors which opens up the possibility of the design and development of robust mimetic agonists and antagonists of natural transmitters.[25, 26] These advances in our understanding of chemical neurotransmission were acknowledged by the award of the 2000 Nobel Prize in Physiology or Medicine to Carlsson, Greengard and Kandel. Carlsson showed how dopamine is involved in the control of movement and opened up the route to our understanding of Parkinson's Disease. Greengard showed how dopamine and similar transmitters work by affecting protein receptors at synapses, while Kandel demonstrated what happens at synapses when learning takes place in a simple system.[27] The editor-in-chief of *Science* perhaps pointed to the main route of future research:

> [The] main research area of interest that will have the greatest impact on chemistry in the future is that of specific neurons and the functional analysis of the circuits in which any given neuron participates.[28]

In short, what is becoming possible is a mechanistic understanding of the biological basis of behaviour.

It is in this context that we should view the increased understanding of muscarinic acetylcholine receptors. The increasing number of known receptor types can be gauged from Table 3, which is taken from a standard listing of receptors and ion channels for the year 2000.[29] This listing gives a total of 44 types of receptor. Table 3 shows some of the better-known examples of these different types of receptor.

TABLE 3
EXAMPLES OF DIFFERENT TYPES OF RECEPTOR

Acetylcholine
Adrenoceptors
Angiotensin
Bombesin
Bradykinin
Cannabinoid
Dopamine
Endothelin
GABA
Glutamate
Glycine
5-HT
Melatonin
Tachykinin

Current Trends. *The 2000 Receptor and Ion Channel Nomenclature Supplement.* London: Elsevier Science, 11th Edn. 2000.

Acetylcholine Receptors

In its overview of work carried out in the 1960s and 1970s, the MoD gives two references of particular interest here. One, on the failure to relate central antimuscarinic potency to behavioural modifications cites a paper by Brimblecombe and Buxton[30] and one on the disruption of conditioned avoidance response linkage to the central antimuscarinic potency in cats cites a paper by Green and Aldous.[31] These papers form part of a collection published in a special issue of *Progress in Brain Research* based on a meeting held at the Chemical Defence Establishment, Porton Down, Salisbury, England in 1971. In their preface to the collection, the editors, P.B. Bradley and R.W. Brimblecombe, noted:

> The fact that more than half the papers were concerned to a greater or lesser extent with cholinergic mechanisms in the central nervous system reflects, in our opinion, the relative importance of acetycholine in brain function.[32]

However, they went on to suggest that growing interest in other transmitter substances had then recently led to the relative neglect of cholinergic mechanisms.

The two papers cited in the MoD overview both used complex measures of behaviour to assess the impact of drugs, but the general approach can be understood from the conclusion of the first paper:

> The basic object of this study was to determine whether the activities of any of these anticholinergic drugs in modifying various aspects of behaviour were correlated with their anticholinergic activities as measured by ability to antagonise oxotremorine-induced salivation or tremors in mice.[30]

As the authors noted at the beginning of their paper, in order to test the anticholinergic properties of a series of drugs they needed a test method for determining central anticholinergic activity and suitable aspects of behaviour to study. In their view at that time:[30]

> The drug oxotremorine seems to lend itself to a test procedure that satisfies the first requirement. This is a muscarinic agonist which produces marked muscular tremors in a variety of species ... These tremors appear to result from stimulation of central muscarinic receptors ... so that the potency of drugs in blocking the tremors gives a measure of their anticholinergic potency.

By exploring this particular issue, we can now understand the difficulties the authors encountered in their work – and the impact of genomics.

Receptor Subtypes

Acetylcholine receptors are divided into two groups: nicotinic acetylcholine receptors and muscarinic acetylcholine receptors. It has long been known that nicotine selectively activates one group of these receptors that naturally respond to acetylcholine whilst another group responds to muscarine, a substance found in the mushroom *Amanita muscaria*.[33] The important point is that there are now known to be some nine subtypes of nicotinic acetylcholine receptor[29] and five subtypes of muscarinic acetylcholine receptor.[29, 34] All the nicotinic ones are directly-acting (ionotropic) receptors whilst all of the muscarinic receptors are indirectly-acting (metabotropic) through G proteins.

Both nicotinic and muscarinic receptors have important roles in peripheral systems, but muscarinic receptors appear to be particularly important in the central nervous system where 'there is evidence that muscarinic receptors are involved in motor control, temperature regulation, cardiovascular regulation and memory'.[34] Not surprisingly, there is intense medical interest in such receptors because, if they were better understood, it might be possible to develop selective therapeutic agents to deal with, for example, Alzheimer's disease and Parkinson's disease.

Muscarinic Receptors

The International Union of Pharmacology's account of muscarinic receptors notes:

> Pharmacological characterization of muscarinic receptor subtypes has long been dogged by a complete lack of agonists with any selectivity, and a lack of antagonists with very high selectivity for any single receptor subtype.[34]

Additionally, cells have been found to express more than one sub-type of these receptors which further adds to the difficulty of defining the function of any single receptor sub-type. Nevertheless, this standard review also noted in 1998 that:

> *Knowledge of the potential functions and roles of muscarinic receptors and their subtypes was advanced significantly by the cloning of five mammalian genes encoding muscarinic receptors ...* Cloning of complementary deoxyribonucleic acids for muscarinic receptor genes was spearheaded by the work of Numa and colleagues, who cloned the M1 and M2 genes, and was extended by the discovery of the M3, M4, and M5 genes [Emphasis added].[34]

As all of these receptor subtypes belong to the G protein-coupled family, a great deal of information is available from knowledge of other such receptors to help elucidate the muscarinic receptor functions.

Furthermore, the impact of genomics has not stopped simply with the cloning of the genes for the receptor sub-types in the 1980s. Advances in gene knock-out technology in the 1990s have allowed further significant progress to be made:

> The availability of the cloned muscarinic receptor genes and recent progress in gene knock-out methodologies have provided the opportunity to examine the physiological roles of the individual muscarinic receptors in an unambiguous fashion.[35]

So it is now possible to alter the relevant receptor gene and make it inoperative in living mice.

A number of such studies have now been carried out and give a picture of the function of the different sub-types of receptor.[36] The receptor relevant to the work carried out on oxotremorine-induced tremors by the MoD in the 1960s turns out to be M2. The function of this subtype of receptor was thoroughly studied in M2 knock-out mice:

> Injection of increasing doses of the centrally acting, nonselective muscarinic agonist OXO [oxotremarine] ... into wild-type control mice reproducibly resulted in massive whole-body tremor ... doses > 0.3 mg/kg resulted in a pronounced increase in lethality.[35]

On the other hand, 'the tremorogenic effects of OXO were absent in M2 –/– mutant mice, even when the OXO dose was increased to 1 mg/kg'. In short, the experiment had a quite clear-cut conclusion.

The authors studied a number of other behavioural consequences of removing the M2 receptor. Some well-known consequences of the injection of oxotremorine were unaffected. For example, the increase in salivary secretion, thought to be mediated by glandular M3 receptors, was the same in wild-type and the M2 mutant mice. The authors stated:

> we have demonstrated that the M2 muscarinic receptor subtype, besides its well documented involvement in controlling cardiac function, plays a key role in mediating muscarinic receptor-dependent movement and temperature control as well as analgesia, three of the most prominent central muscarinic effects.

They ended by arguing that these findings 'should provide a rational basis for the development of novel muscarinic drugs'. They reinforced this point strongly in a second study of M4 knock-out mice.[37] These receptors were not implicated in the functions elucidated for the M2 receptor, but rather in the control of basal locomotor activity together with D1 dopamine receptors (Table 4).

This clearly opens up the possibility of finding drugs to combat Parkinson's Disease without the well known side-effects of currently available drugs. Rapid further developments in our understanding of

TABLE 4
SUMMARY OF SOME MUSCARINIC RECEPTOR SUBTYPE FUNCTIONS

Genetic model	Functional responses to muscarinic agonist	Other phenotypic manifestations
M2 muscarinic receptor knockout mice	Disrupted OXO-induced tremor Diminished OXO-induced analgesia Attenuated OXO-induced hypothermia Altered muscarinic regulation of heart rate Preserved OXO-induced salivation	
M4 muscarinic receptor knockout mice	Preserved hypothermia, salivation, tremor, analgesia, induced by OXO	Hyperactivity and D1 dopamine receptor supersensitivity

RP Gainetdinov, MG Caron. Delineating Muscarinic Receptor Functions. *Proc Natl Acad Sci* 1999; 96: 12222–3.

muscarinic receptor subtypes is therefore to be expected.[38] Whilst use of a chemical specifically designed to affect M2 receptors would still cause a variety of behavioural consequences, it is not impossible that further differentiation between M2 receptors could be achieved.[18]

Adrenergic Receptors

During the 1960s and 1970s many glycollates were investigated for their psychoactive effects.[39] Glycollates other than just BZ could have presented a threat as incapacitants.[4] It is important to grasp, however, that it was not just disruption of acetylcholine receptors that was being investigated. There was clearly interest from the early post-war years, in the major US research programme, in means of interfering with the adrenergic systems of the brain.[15] This programme continued into the 1990s and a paper at the 1995 Edgewood Annual Scientific Conference, summarizing 40 years of work there, stated:

> Depending on the specific scenario, several classes of chemicals have potential use, to include: potent analgesics; anaesthetics as rapid acting immobilisers; sedatives as immobilisers; and calmatives that leave the subject awake and mobile but without the will or ability to meet objectives.[40]

There was certainly evidence that some of the chemicals being investigated would have interfered with adrenergic systems.[15]

Adrenergic neurones secrete adrenaline or, more commonly, noradrenaline as a neurotransmitter. Such neurones occur in both the peripheral and the central nervous systems. One specialized sub-system of the brain is called the *locus coeruleus*. It consists of a relatively small number of neurons, some 20,000, but most of these secrete noradrenaline. Furthermore, very few central noradrenaline neuronal cell bodies are located outside of this structure. What is surprising is that the axons from these cells ramify very widely throughout the brain. The sub-system appears to play a role in arousal and attention. The Edgewood Conference of 1994 had a paper that stated:

> Centrally acting α_2-adrenergic compounds show antihypertensive actions with sedative properties. More selective α_2-adrenergic compounds with potent sedative activity have been considered to be ideal next generation anesthetic agents which can be developed and used in the Less-Than-Lethal technology program.[41]

There are known to be three distinct types of receptor for adrenaline and noradrenaline. These are termed the α_1, α_2 and β adrenoceptors (or adrenergic receptors) and each type has three sub-types, making a total of nine potential targets for drugs or chemical agents.[29]

New agents were being sought in the early 1990s to specifically affect some of the activities controlled by central adrenergic systems. Another paper at the 1994 Edgewood Conference stated:

> Medetomidine represents a new class of α_2-adrenergic agonists. Both sedative and hypotensive effects are noted with medetomidine. As part of a search for a new agent of the medetomidine series which shows potent anaesthetic, analgesic and sedative activity with a lack of adverse cardiovascular side effects we discovered that a naphthalene analogue of medetomidine has greater potency as well as higher selectivity for α_2-adrenergic receptors.[42]

It was not possible, however, to follow this research further in the open literature.

It is known that adrenoceptors are involved in diverse physiological functions, particularly of the central nervous system and the cardiovascular system and adrenoreceptor agonists are used in the treatment of hypertension, glaucoma, attention-deficit disorder, etc. However, it was difficult to assign specific functions to specific sub-types because of the lack of sufficiently selective agonists and antagonists. As with the acetylcholine system, it has nevertheless been possible to elucidate the functions of sub-types through the use of knockout mice.[43]

In regard to the potential use of chemical agents for sedation, 'α–adrenergic agonists appear to induce sedation by activating autoreceptors in the locus coeruleus reducing its spontaneous rate of firing'.[43] So the chemical agonist causes inhibition of the locus coeruleus noradrenaline neurones by activating inhibitory synapses on these cells which would normally be affected by the noradrenaline they usually produce (i.e., autoinhibition). This natural damping effect is thus accentuated and output of noradrenaline in other parts of the brain innervated by processes from these neurones is reduced.

Animals with a very small mutation in the α_{2A} sub-type of adrenoceptor (which effectively disrupts its function and is thus equivalent to a complete gene knockout) were not affected by the sedative dexmedetomidine,[44, 45] neither were neurones from the locus coeruleus of these animals affected by this agent.[45] Mice made without the other sub-types of adrenoceptor (α_{2B}, α_{2C}) by genetic engineering were affected in the same way as normal mice, strongly supporting the view that the α_{2A} sub-type is the receptor sub-type of importance in sedation. This opens the way for more selective agents to be developed.

Conclusion

During the 1960s and 1970s, prior to the genomics revolution, many glycollates were investigated for their psychoactive effects.[39] Moreover, others than just BZ could have presented a threat as incapacitants.[4] The problem for a weapons designer at that time would have been the inability to obtain specific controllable effects, because the targets for which chemicals were being designed had not been properly characterized. In the unlikely event that still holds true today, it is an open question for how long it will last given the pace and direction of legitimate work.

Should *undesirable* effects be obtainable by action on muscarinic or any of the many other potential receptor sub-types,[29] the task of preventing large-scale misuse of this knowledge could be very difficult, given the loophole in the CWC. Careful attention needs to be paid now to the actions that might be taken, inside or outside of the CWC, to avoid going down this road.

Notes

An earlier version of this paper was presented under Agenda Item II 2a (Advances in Science and Technology) to the15th Workshop of the Pugwash Study Group on the Implementation of the Chemical and Biological Weapons Conventions: 'Approaching the First CWC Review Conference', Oegstgeest, The Netherlands, 23–24 June 2001. The work was financed in part by grant No.054752 from the Wellcome Trust.

1. Robertson G. Oral Answers to Questions. *Hansard* 9 Feb. 1998 (Pt.1) Col.3. London: House of Commons, 1998.
2. Ministry of Defence. *Iraq CW Capability during the Gulf War: Agent 15.* London: Ministry of Defence, 1998.
3. Ketchum, JS, *et al.* Incapacitating Agents. In: Sidell, FR, *et al.*, eds. *Medical Aspects of Chemical and Biological Warfare.* Washington DC: Office of the Surgeon General, Department of the Army, 1997: 287–306.
4. DERA. *An Overview of Research Carried out on Glycollates and Related Compounds at CBD Porton Down.* DERA/CBD/CR 990418, Sept. 1999.
5. Bowman, S. *Iraqi Chemical and Biological Weapons (CBW) Capabilities.* CRS Issue Brief, April 1998. Washington DC: Congressional Research Service, 1998.

6. Anon. Iraq 'horror weapon' warning. *Yorkshire Post*, 10 Feb. 1998: 5.
7. Brown C, Burrell I. Iraqi 'zombie gas' arsenal revealed. *Independent*, 10 Feb. 1998: 9.
8. Butcher T. Saddam has 'Agent 15' knock-out gas in store. *Daily Telegraph*, 10 Feb. 1998: 15.
9. Krutzsch W, Trapp R. *A Commentary on the Chemical Weapons Convention*. Dordrecht: Martinus Nijhoff, 1994.
10. Editorial. New Technologies and the Loophole in the Convention. *Chemical Weapons Convention Bulletin*, March 1994; 23: 1–2.
11. Meselson MS. *Statement to Hearing on Chemical Weapons Ban Negotiations Issues*. Committee on Foreign Relations: US Senate, 1 May 1992: 23–5.
12. Anon. Chronology. *Chemical Weapons Convention Bulletin*, June 1994; 25: 23.
13. Perry Robinson J. The 1993 Chemical Weapons Convention. Paper prepared for the Harvard Sussex Programme London CBW Seminar, 10 Jan. 1993.
14. Woodhouse T, Bruce R, Dando MR. *Peacekeeping and Peacemaking: Towards Effective Intervention in Post-Cold War Conflicts*. London: Macmillan, 1997.
15. Dando MR. *A New Form of Warfare: The Rise of Non-Lethal Weapons*. London: Brassey's, 1996.
16. Lewer N, Feakin T, Dando MR. The Future of Non-Lethal Weapons. *Med Confl Surviv* 2001; 17 (3): 175–9.
17. Edwards R. War without Tears. *New Sci* 9 Dec. 2000: 3–4.
18. Burgen ASV. Targets of Drug Action. *Annu Rev Pharmacol Toxicol* 2000; 40: 1–16.
19. Kandel ER, Squire LR. Neuroscience: Breaking Down Scientific Barriers to the Study of Brain and Mind. *Science* 2000; 290: 1113–29.
20. Birdsall NJM, et al. Muscarinic Receptors: It's a Knockout. *Trends in Pharmacological Sciences* 2001; 22 (5): 215–19.
21. Lander ES, Weinberg RA. Genomics: Journey to the Center of Biology. *Science* 2000; 287: 1777–82.
22. Rose S. The Future of the Brain. *Biologist* 2000; 47 (2): 96–9.
23. Levitan IB, Kaczmarek LK. *The Neuron: Cell and Molecular Biology*. Oxford: Oxford University Press, 1997.
24. Dando MR. Genomics, Bioregulators, Cell Receptor Research and Potential Biological Weapons: Considerations Regarding the Scope of Article I of the Biological and Toxin Weapons Convention (BTWC). Geneva: Paper presented at Pugwash Meeting No.258, 'Key Issues for the Fifth BWC Review Conference 2001', 18–19 Nov. 2000.
25. Dando MR. Receptors and Ligands: Applications of Particular Relevance in Relation to the Possible Misuse of Neuroscience. Prague: Paper presented at a NATO Advanced Research Workshop, 'New Scientific and Technological Developments of Relevance to the Biological and Toxin Weapons Convention', 31 May–2 June 2001.
26. Dando MR. *The New Biological Weapons: Threat, Proliferation and Control*. Boulder, CO: Lynne Rienner, 2001.
27. Marchant J. The Prozac Generation: A Trio of Brain Researchers Wins This Year's Nobel Prize for Medicine. *New Sci*, 14 Oct. 2000: 7.
28. Bloom FE. Special Report: Chemistry in the Service of Humanity. *Chemical & Engineering News*, 6 Dec. 1999: 66.
29. Current Trends. *The 2000 Receptor and Ion Channel Nomenclature Supplement*. London: Elsevier Science, 11th Edn. 2000.
30. Brimblecombe RW, Buxton DA. Behavioural Actions of Anticholinergic Drugs. *Prog Brain Res* 1972; 36: 115–26.
31. Green DM, Aldous FAB. The Effects of Anticholinergic Drugs, Chlorpromazine and LSD-25 on Evoked Potentials, EEG and Behaviour. *Prog Brain Res* 1972; 36: 143–58.
32. Bradley PB, Brimblecombe RW. Preface: Biochemical and Pharmacological Mechanisms Underlying Behaviour. *Prog Brain Res* 1972; 36: x.
33. Conn PM. *Neuroscience in Medicine*. Philadelphia: JB Lippincott, 1995.
34. Caulfield MP, Birdsall NJM. International Union of Pharmacology. XVII. Classification of Muscarinic Acetylcholine Receptors. *Pharmacological Reviews* 1998; 50 (2): 279–90.
35. Gomez J, et al. Pronounced Pharmacologic Deficits in M2 Muscarinic Acetylcholine Receptor Knockout Mice. *Proc Natl Acad Sci* 1999; 96: 1692–7.
36. Gainetdinov RP, Caron MG. Delineating Muscarinic Receptor Functions. *Proc Natl Acad Sci* 1999; 96: 12222–3.
37. Gomez J, et al. Enhancement of D1 Dopamine Receptor-Mediated Locomotor

Stimulation in M4 Muscarinic Acetylcholine Receptor Knockout Mice. *Proc Natl Acad Sci* 1999; **96**: 10483–8.

38. Nathanson NM. A Multiplicity of Muscarinic Mechanisms: Enough Signalling Pathways to Take your Breath Away. *Proc Natl Acad Sci* 2000; **97**: 6245–7.

39. Perry Robinson J. The Chemical Industry and Chemical-Warfare Disamament: Categorizing Chemicals for the Purposes of the Projected Chemical Weapons Convention. In: *SIPRI Chemical and Biological Warfare Studies No.4*. Stockholm: Stockholm International Peace Research Institute, 1986; 55–104.

40. Anon. *Chemical Weapons Convention Bulletin*, March 1996; **27**: 14 (entry dated 14–17 Nov. 1995).

41. Hsu F-L, Przeslowski RM. Synthesis and α_2-Adrenergic Activity of Quinoline and Quinoxaline Analogues of Medetomidine. In: *Scientific Conference on Chemical and Biological Defense Research*. Maryland: Edgewood Research, Development and Engineering Center, Aberdeen Proving Ground, 1994: 18.

42. Miller DD, *et al*. Synthesis and Biological Activity of a Series of Conformationally Restricted Analogues: 4-Substituted Imidazoles as α_2-Adrenergic Agonists. In: *Scientific Conference on Chemical and Biological Defense Research*. Maryland: Edgewood Research, Development and Engineering Center, Aberdeen Proving Ground, 1994: 17.

43. Kable JW, *et al*. In Vivo Gene Modification Elucidates Subtype-Specific Functions of α_2-Adrenergic Receptors. *J Pharmacol Exp Ther* 2000; **293**: 1–7.

44. Hunter JC, *et al*. Assessment of the Role of α_2-Adrenoceptor Subtypes in the Antinociceptive, Sedative and Hypothermic Action of Dexmedetomidine in Transgenic Mice. *Br J Pharmacol* 1997; **122**: 1339–44.

45. Lakhlani PP, *et al*. Substitution of a Mutant α_2-Adrenergic Receptor via Hit and Run' Gene Targeting Reveals the Role of this Subtype in Sedative, Analgesic, and Anaesthetic-Sparing Responses *in vivo*. *Proc Natl Acad Sci* 1997; **94**: 9950–55.

Select Bibliography

This bibliography reflects the eclectic nature of the subject of non-lethality, which engages a wide range of issues including: weapons technology, military strategy and doctrine, ethics, humanitarian intervention and peace support operations, civilian policing, human rights, international humanitarian law, the laws of war and international arms control legislation. Readers are also referred to more extensive bibliographies. These include: Joan Hyatt (1995), Robert Bunker (1996), the extensive bibliography produced by Steve Wright of the Omega Foundation for the Scientific and Technical Options Assessment Research Division (STOA) of the European Parliament (1998) and Nick Lewer (2000).

Ackroyd C, Margolis K, Shallice T. *The Technology of Political Control*. Harmondsworth: Pelican, 1977.

Aftergood S. The Soft-Kill Fallacy. *Bull Atom Sci* 1994; **50** (5): 44–5.

Alexander J. *Future War: Non-Lethal Weapons In 21st Century Warfare*. New York: St Martin's Press, 1999.

Altmann J. *Acoustic Weapons – A Prospective Assessment: Propagation, and Effects of Strong Sound*. Ithaca, NY: Cornell University Peace Studies Program, 1999.

Altmann J. Acoustic Weapons: A Prospective Assessment. *Science and Global Security* 2001; **9**: 49–121.

Amnesty International. *Arming the Torturers: Electro-shock Torture and the Spread of Stun Technology*. New York: AI, 1997.

Amnesty International. *United States of America: Use of Electro-shock Stun Belts*. New York: AI, 1996.

Anderberg B, Wolbarsht M. Blinding Lasers: The Nastiest Weapon? *Military Technology*, March 1990.

Applegate R. *Riot Control – Materiel and Techniques*. Harrisburg: Stackpole Books, 1969.

Applegate R. Nonlethal Police Weapons. *Ordnance*, July–Aug. 1971: 62–6.

Arkin W. Acoustic Anti-PersonnelWeapons: An Inhumane Future? *Med Confl Surviv* 1997; **14**: 314–26.

Bailey W. Less-than-Lethal Weapons and Police-Citizen Killings in U.S. Urban Areas. *Crime and Delinquency* 1996; **42** (4): 535–52.

Barry J, Everett M, Peck A. Nonlethal Military Means: New Leverage for a New Era. National Security Program Policy Analysis Paper 94-101, John F Kennedy School of Government, Harvard University, 1994.

Black S. Non-Lethal Weapons Systems: The Potential Impact of New Technologies on Low Intensity Conflict. Matthew B Ridgeway Center for International Security Studies, University of Pittsburgh, Graduate School of Public and International Affairs, Ridgeway Viewpoints, No.93–9, 1993.

Bunker RJ, Moore TL. Nonlethal Technology and Fourth Epoch War: A New Paradigm of Politico-Military Force. Land Warfare Paper No.23. Arlington, VA: Institute of Land Warfare, Association of the United States Army, 1996.

Bunker RJ, ed. *Nonlethal Weapons: Terms and References*. INSS Occasional Paper 15. Washington, DC: USAF Institute for National Security Studies, 1996.

Carnegie Commission on Preventing Deadly Conflict. *Preventing Deadly Conflict. Final Report*. New York: Carnegie Corporation of New York, 1997.

Center for Strategic and International Studies (CSIS). *The Role of NL Capabilities in Limiting Effects on Interstate and Intrastate Violence*. Washington, DC: CSIS, 1995.

Chemical Weapons Convention – 1993. In: *International Legal Materials* 1993; 32: 800–73.

Coates JF. Non-Lethal Police Weapons. *Technology Review* June 1972: 49–56.

Coates JF. *Nonlethal and Nondestructive Combat in Cities Overseas. Paper P-569*. Arlington, VA: Institute for Defense Analyses, Science and Technology Division, May 1970.

Collins JM. *Non-Lethal Weapons and Operations: Potential Applications and Practical Limitations*. Washington, DC: Congressional Research Service Report for Congress, Library of Congress, Sept. 1995.

Coppernoll M-A, Maruyama, X. Legal and Ethical Guiding Principles and Constraints Concerning NLWs Technology and Employment. Paper presented at Non-Lethal Defence III Conference, John Hopkins Applied Physics Laboratory, 25–26 Feb. 1998.

Coppernoll MA. The Non-Lethal Weapons Debate. *Naval War College Review* 1999; 52: 112–31.

Council for Foreign Relations. *Non-Lethal Technologies: Military Options and Implications*. Washington, DC: Report of an Independent Task Force, Council for Foreign Relations, 1995.

Council on Foreign Relations. *Independent Task Force Report – Nonlethal Technologies: Progress and Prospects*, 1999, http://www.foreignrelations.org/public/pubs/Non-ViolentTaskForce.htm.

Coupland R. 'Non-Lethal' Weapons: Precipitating a New Arms Race. *BMJ* July 1997; **315**: 72.

Coupland R. *Criteria for Judging Excessively Harmful Weapons and Weapons Which Cause 'Superfluous Injury and Unnecessary Suffering'*. Geneva: SIrUS Project – Health Division, ICRC, April 1997.

Curtis L. *They Shoot Children: The Use Of Rubber and Plastic Bullets in the North of Ireland*. London: Information on Ireland, 2nd Edn. 1987.

Dando M. *A New Form of Warfare: The Rise of Non-Lethal Weapons*. London: Brassey's, 1996.

Dando M, ed. *Non-lethal Weapons: Technological and Operational Prospects*. Coulsdon: Jane's, 2000.

Dando MR. *The New Biological Weapons: Threat, Proliferation and Control*. Boulder, CO: Lynne Rienner, 2001.

Deane-Drummond, A. *Riot Control*. London: Royal United Services Institute for Defence Studies, 1975.

Dewar M. *Weapons and Equipment of Counter-Terrorism*. London: Arms and Armour Press, 1995.

Doswald-Beck L, ed. *Blinding Weapons: Reports of the Meetings of Experts Convened by the International Committee of the Red Cross on Battlefield Laser Weapons, 1989–1991*. Geneva: International Committee of the Red Cross, 1993.

Doswald-Beck L. New Protocol on Blinding Laser Weapons. *International Review of the Red Cross* May–June 1996; **312**: 272–99.

Egner DO. *The Evolution of Less-Lethal Weapons* (Microfiche). Aberdeen Proving Grounds, US: Human Engineering Laboratory, Dec. 1977.

Ezell E. *Small Arms Today*. Boston: Stackpole, 1988.

Faul D, Murray R. *Rubber and Plastic Bullets Kill and Maim*. Belfast: Association for Legal Justice, 1981.

Fessler A. Benign Intervention – An Idea Whose Time has Come? *Friend's Quarterly* 1995; **29** (6): 246–55.

Fidler DP. The International Legal Implications of 'Non-Lethal' Weapons. *Michigan Journal of International Law* 1999; **21**: 51–100.

Foster G. Nonlethality: Arming the Postmodern Military. *RUSI Journal* Oct. 1997: 56–63.

Freedman L. *The Revolution in Military Affairs*. London: International Institute for Strategic Studies/Oxford University Press, 1998: 45.

Frost G, Shipbaugh C. *GPS Targetting Methods For Non-Lethal Systems*. Santa Monica, CA: Rand Publication RP-262, 1994.

Geneva Protocol, 1925. In: Roberts A, Guelff R, eds. *Documents on the Laws of War*. Oxford: Oxford University Press, 2nd Edn. 1989: 139–40.

Gillow T. The Psychological, Social and Economic Consequences of Blinding Soldiers. *Med Confl Surv* 1997; 13 (4): 327–32.

Girard, H. Non-Lethal Weapons Policy: The Case of Electromagnetic Weapons. Paper presented at Annual Meeting of the Canadian Association of Security and Intelligence Studies, University of Quebec, 5 June 1995.

Goldblat J. Inhumane Conventional Weapons: Efforts to Strengthen Constraints. *Arms, Disarmament and International Security*, Stockholm: SIPRI, 1995.

Goolsby T. Aqueous Foam as a Less-than-Lethal Technology for Prison Applications. *SPIE-International Society for Optical Engineering* 1997; 2934: 86–95.

Gurr NJ. Non-Lethal Weapons: Lifting the Lid on the Issue of Human Rights. *International Journal of Human Rights* 1997; 1 (4): 1–17.

High Commissioner for Human Rights. *Human Rights and Law Enforcement – A Manual on Human Rights Training for the Police*. Professional Training Series No.5. Geneva: United Nations, 1997.

Hiss J, Hellman F, Kahana T. Rubber and Plastic Ammunition Lethal Injuries: The Israeli Experience. *Medicine, Science and the Law* 1997; 37 (Part 2): 139–44.

Howard M. The Forgotten Dimensions of Strategy. *Foreign Affairs* 1979; 57: 976–86.

Independent Commission on Policing in Northern Ireland. *A New Beginning: Policing in Northern Ireland*. London: HMSO, 1999.

Human Rights Arms Watch Project. *Blinding Laser Weapons: The Need To Ban a Cruel and Inhumane Weapon*. Washington, DC: Human Rights Watch, Sept. 1995.

Human Rights Watch Arms Project. US Blinding Laser Weapons. *Human Rights Arms Watch* May 1995; 7 (5).

Hyatt J. *Nonlethal Weapons Bibliography*. Maxwell AFB, Al, US: Air University Library, June 1995.

Jussila J. Wounding Potential of Handgun Bullets – The Relative Damage Index. *Annales Medicine Militaris Fenniae* 2000; 75: 125–34.

Kalshoven F. *Constraints on the Waging of War*. Geneva, International Committee of the Red Cross, 1987.

Kalshoven F. The Conventional Weapons Convention: Underlying Legal Principles. *International Review of the Red Cross* 1990; **279**: 510–20.

Ketchum, JS, *et al.* Incapacitating Agents. In: Sidell FR, *et al.* eds. *Medical Aspects of Chemical and Biological Warfare*. Washington DC: Office of the Surgeon General, Department of the Army, 1997: 287–306.

Kock E, Rix B. *A Review of Police Trials of the CS Aerosol Incapacitant*. Police Research Series, Paper 21. London: Home Office, Police Research Group, 1996.

Klaaren, JW, Mitchell RA. Nonlethal Technology and Airpower: A Winning Combination for Strategic Paralysis. *Airpower Journal* 1995: 42–51.

Knoth A. Disabling Technologies: A Critical Assessment. *International Defense Review* July 1994: 30–39.

Kokoski R. Non-Lethal Weapons: A Case Study of New Technology Developments. In: *SIPRI Yearbook 1994*. Oxford: SIPRI/Oxford University Press, 1994: 367–86.

Land Mine Convention, 1997. In: *International Review of the Red Cross* 1997; **320**: 563–78.

Lewer N. Non-Lethal Weapons. *Med War* 1995; **11**: 78–95.

Lewer N, Schofield S. *Non-Lethal Weapons: A Fatal Attraction?* London: Zed Books, 1997.

Lewer N, Schofield S. Non-Lethal Weapons for U.N. Military Operations. *International Peacekeeping* 1997; **4** (3): 71–93.

Lewer N. Bibliography of Non-Lethal Weapons. In: Dando M, ed. *Non-lethal Weapons: Technological and Operational Prospects*. Coulsdon: Jane's, 2000: 88–133.

Lewer N. Benign Intervention and Non-Lethality: Wishful Thinking For The 21st Century. In: Dando M, ed. *Non-lethal Weapons: Technological and Operational Prospects*. Coulsdon: Jane's, 2000: 67–77.

Lorenz FM. Less-Lethal Force in Operation United Shield. *Marine Corps Gazette* Sept. 1995: 69–76.

Lorenz FM. Non-Lethal Force: The Slippery Slope to War. *Parameters* Autumn 1996: 52–62.

Lovelace D, Metz S. *Nonlethality and American Land Power: Strategic Context and Operational Concepts*. Carlisle, US: Army War College, 1998.

Lyell Lord. Non-Lethal Weapons. *Draft General Report*. North Atlantic Assembly, Science and Technology Committee, AP238, STC (97) 8, Sept. 1997.

Mazarr MJ. *The Military Technological Revolution: A Structural Framework*. Washington, DC: Center for Strategic and International Studies, March 1993.

Metz S, Kievit J. *The Revolution in Military Affairs and Conflict Short of War.* Carlisle: US Army War College, 1994.

Metz S. The Next Twist of the RMA. *Parameters* 2000; **XXX** (3): 40–53.

Metz S. Which Army After Next? The Strategic Implications of Alternative Futures. *Parameters* Autumn 1997: 15–26.

Metz S. *Armed Conflict in the 21st Century: The Information Revolution and Post-Modern Warfare.* Carlisle: US Army War College, 2000.

Millar R, Rutherford W, *et al.* Injuries Caused by Rubber Bullets: A Report on 90 Patients. *Brit J Surg* 1975; **62**: 480–86.

Morehouse D. *Nonlethal Weapons: War Without Death.* Westport: Praeger, 1996.

Morris C, Morris J. End Battle over Nonlethals. *Defense News* 13–19 Nov. 1995: 40.

Morris C, Morris J, Baines T. Weapons of Mass Protection: Nonlethality, Information Warfare, and Airpower in the Age of Chaos. *Airpower Journal* 1995: 15–29.

Morris J, Krivorotov V, Morris C. *The Age of Chaos: Threat and Solution Beyond Containment. A Framework For Nonlethal Peacemaking and Peacekeeping,* Box 312, West Hyannisport, MA 02672, 1993.

O'Connell E, Dillapain J. Nonlethal Concepts: Implications For Air Force Intelligence. *Airpower Journal* 1994; **8** (Part 4): 26–33.

Ordog GJ, Wasserberger J, *et al.* Electronic Gun (Taser) Injuries. *Ann Emerg Med* 1987; **16** (1): 73–8.

Perry Robinson J. The Chemical Industry and Chemical-Warfare Disamament: Categorizing Chemicals for the Purposes of the Projected Chemical Weapons Convention. In: *SIPRI Chemical and Biological Warfare Studies No.4.* Stockholm: Stockholm International Peace Research Institute: 1986; 55–104.

Perry Robinson JP. Developments in 'Non-Lethal Weapons' Involving Chemicals. In: *Expert Meeting on Certain Weapon Systems and on Implementation Mechanisms in International Law. Report of Meeting, Geneva 30 May–1 June 1994.* Geneva: International Committee of the Red Cross, 1994.

Peters A. Blinding Laser Weapons. *Med Confl Surv* 1996; **12** (2): 107–13.

Prokosch E. *The Technology of Killing: A Military and Political History of Anti-Personnel Weapons.* London: Zed Books, 1995.

Roberts A, Guelff R, eds. *Documents on the Laws of War.* Oxford: Oxford University Press, 2nd Edn. 1989.

Rogers P. *Losing Control.* London: Pluto Press, 2000.

Sapolsky H. *Non-Lethal Warfare Technologies: Opportunities and Problems.* Massachusetts: MIT Defense and Arms Control Studies Programme, 1993.

Security Planning Corporation. *Nonlethal Weapons for Law Enforcement: Research Needs and Priorities.* Washington, DC: Security Planning Corporation, 1972.

Shaw J. Pulmonary Contusion in Children due to Rubber Bullet Injuries. *BMJ* 1972; 4: 764–6.

SIPRI. Electric, Acoustic and Electromagnetic-wave Weapons. In: *SIPRI Yearbook 1978.* London: Taylor & Francis, 1978: Ch.8.

SIPRI. The Prohibition of Inhumane and Indiscriminate Weapons. In: *World Armaments and Disarmament.* Oxford: Oxford University Press, 1981: Ch.15.

SIPRI. *The Law of War and Dubious Weapons.* Stockholm: SIPRI, 1976.

Sollenburg M, Wallensteen P. Global Patterns of Major Armed Conflicts, 1989–1996. In: *SIPRI Yearbook 1997: Armaments, Disarmament and International Security.* Oxford: Oxford University Press, 1998: 17.

Sweetman S. *Report on the Attorney General's Conference on Less Than Lethal Weapons.* US Department of Justice/National Institute of Justice: Office of Communication and Research Utilization, 1987.

Toffler A, Toffler H. *War and Anti-War: Survival At The Dawn Of The 21st Century.* Boston, MA: Little, Brown, 1993; Ch.15.

Training and Doctrine Command (TRADOC). *Military Operations. Concept For Non-Lethal Capabilities In Army Operations.* Fort Munroe, VA: TRADOC Pamphlet 525-73, Department of Army, 1 Sept. 1996.

Training and Doctrine Command (TRADOC). *Concept For Nonlethal Capabilities In Army Operations.* Fort Munroe, VA: TRADOC, 1 Dec. 1996.

Truesdell A. *The Ethics of Non-Lethal Weapons.* Occasional Paper No.24. Camberley: Staff College, Strategic and Combat Studies Institute, 1996.

Vesaluoma M, Muller L, *et al.* Effects of Oleoresin Capsicum Pepper Spray on Human Corneal Morphology and Sensitivity. *Invest Opthalmol Vis Sci,* 2000; 41: 2138–47.

Walzer M. *Just and Unjust Wars.* New York: Basic Books, 1977.

Woodhouse T, Bruce R, Dando MR. *Peacekeeping and Peacemaking: Towards Effective Intervention in Post-Cold War Conflicts.* London: Macmillan, 1997.

Wright S. *An Appraisal of Technologies of Political Control – A Report for the European Parliament.* Luxembourg: European Parliament Scientific and Technological Options Assessment (STOA), 1998: 166–499.

Notes on Contributors

John B. Alexander is one of the world's leading experts on non-lethal weapons. As a Colonel in the US Army, his extensive military experience included commanding Green Berets in Vietnam. He was a Deputy-Sheriff in Dade County, Florida, for five years. While working at Los Alamos National Laboratory he developed the concept of 'Non-Lethal Defense' and was engaged in advising the US military to adopt a coherent policy towards non-lethal weapons. He is currently the Science Director for the Appollinaire Group (a private research organization) in Las Vegas and a consultant to the US Government. His many publications include *Future War: Non-Lethal Weapons in 21st Century Warfare* (1999).

Jürgen Altmann is a physicist specializing in scientific and technical aspects of disarmament. In 1988 he founded the Bochum Verification Project which studies the potential of automatic sensor systems for co-operative verification of disarmament and peace agreements. He is also interested in new military technologies, particularly their prospective assessment and preventive limitation. He has written widely on the subjects of beam weapons, non-lethal (in particular, acoustic) weapons and military applications of micro-system technologies.

Colin Burrows is a Chief Superintendent in the Royal Ulster Constabulary, responsible for the development of Operational Policy and Support. He holds the Queen's Police Medal.

Malcolm Dando is Professor of International Security at the Department of Peace Studies, University of Bradford. A biologist by training, he has published extensively in the area of non-lethal weapons, biological weapons and arms control. His recent books include: *A New Form of Warfare: The Rise of Non-Lethal Weapons* (1996); *Non-Lethal Weapons: Technological and Operational Prospects* (2000); *The New Biological Weapons: Threat, Proliferation and Control* (2001).

Tobias Feakin is a Ph.D. researcher at the Department of Peace Studies, University of Bradford and is Research Associate of the Bradford Non-Lethal Weapons research project. His current work focuses on the use of non-lethal weapons in South Asia, with particular emphasis on India. His previous work examined the use of non-lethal weapons by UK Police Forces.

David P. Fidler is Professor of Law at Indiana University School of Law. He is one of the world's leading experts on international law and public health, with an emphasis on infectious diseases. His many publications include: *International Law and Public Health: Materials On and Analysis of Global Health Jurisprudence* (2000) and *International Law and Infectious Diseases* (1999). He has served as an international legal consultant to the World Health Organization, the Center for Disease Control and Prevention, the US Department of Defense, the World Bank Group and the Federation of American Scientists.

Jorma Jussila is Senior Advisor in Weapons Technology at the Police Technical Centre, Helsinki, Finland. He holds an MS in Information Technology. He has previously worked as a consultant systems analyst with the Sperry Corporation, as a Chief Inspector in the Security Service of Finland and as a Senior Advisor of Information Security at the Helsinki Police Department. He is the author of *Handgun in Professional Use* (1997 – in Finnish).

Nick Lewer is Senior Lecturer and Director of the Centre for Conflict Resolution, Department of Peace Studies, University of Bradford, specializing in mediation and peacebuilding in situations of violence and in non-lethal weapons. His publications include (with Oliver Ramsbotham) *'Something Must Be Done': Towards an Ethical Framework for Humanitarian Intervention in International Social Conflict* (1993); and (with Steven Schofield) *Non-Lethal Weapons: A Fatal Attraction? Military Strategies and Technologies for 21st Century Conflict* (1997).

Brian Rappert is a Post-Doctoral Fellow in the School of Sociology and Social Policy at the University of Nottingham. He has published numerous articles on non-lethal weapons and also regarding the sociology of technology, scientific and technical professions and public policy. A central concern to his work has been the consideration of how organizations can, and do, make decisions about technologies in situations of uncertainty and disagreement. He is the author of *Technology, Politics and Conflict: Non-Lethal Weapons as Legitimating Forces?* (forthcoming).

Gerrard Quille is Deputy Director of the International Security Information Service (ISIS), where he was previously Research Officer examining the 1998 Strategic Defence Review within the context of the 'Revolution in Military Affairs'. Until recently he was a Ph.D. researcher at the Department of Peace Studies, University of Bradford, focusing on the Strategic Defence Review (SDR). His publications include the ISIS Briefing Papers *NATO Enlargement* (Paper No.67, 1997) and *The Revolution in Military Affairs and the UK* (Paper No.70, 1998).

Wing Commander Victor Wallace (RAF) qualified in medicine from Aberdeen University in 1984. Following training in general practice he joined the Royal Air Force in 1990. In 1999, he attended No.3 Advanced Command and Staff Course at the Joint Services Command and Staff College during which the chapter in this book was submitted as his Defence Research Paper. Wing Commander Wallace is currently Chief Flight Medicine at the United States Air Force Medical Operations Agency, Headquarters USAF Surgeon General, Washington, DC.

Steve Wright has a B.Sc. in Science and Technology Policy from the University of Manchester and a Ph.D. from the University of Lancaster. He headed Manchester City Council's Police Monitoring Unit from 1984 and is an advisor to Amnesty International in their work on the proliferation on electro-shock and stun technologies. He has worked with many non-government organizations examining the role of internal security technology in relation to human rights violations, including Oxfam's recent project 'Small Arms – Wrong Hands'. In 1990, he helped establish the Omega Foundation which set out to track the international trade in military, security and police technologies and particularly their abuses. His many publications include (for the European Parliament) *An Appraisal of the Technologies of Political Control* (1998) and *Crowd Control Technologies* (2000).

Index